STATISTICS, SCIENCE AND PUBLIC POLICY

Proceedings of the Conference on Statistics, Science and Public Policy held at Herstmonceux Castle, Hailsham, U.K. April 10-13,1996

A.M. HERZBERG *and* I. KRUPKA *Editors*

Canadian Cataloguing in Publication Data

Conference on Statistics, Science and Public Policy (1996 : Hailsham, England)
Statistics, science and public policy : proceedings of the Conference on Statistics, Science and Public Policy held at Herstmonceux Castle, Hailsham, U.K., April 10-13, 1996

Includes bibliographical references.
ISBN 0-88911-798-5

1. Science and state – Congresses. 2. Statistics – Congresses. 3. Political planning – Congresses. I. Herzberg, A.M. II. Krupka, I. (Ivo). III. Title.

Q124.6.C663 1996 338.9'26 C98-930622-4

Queen's University, Kingston, Ontario, Canada K7L 3N6

Table of Contents

Part III: PUBLIC POLICY

Part IV: HAZARDS AND RISKS

Part V: STATISTICS

APPENDIX

Foreword

These proceedings summarize the presentations at the Conference on Statistics, Science and Public Policy held at Herstmonceux Castle in the United Kingdom from April 10-13, 1996. A summary of the discussion at the opening session has also been included.

Since the conference was aimed at promoting dialogue, participants who made presentations were free to choose their format. Some presented formal papers, others gave more informal presentations, and one or two spoke extemporaneously, adjusting their remarks in the light of themes explored by previous speakers.

In the interests of space, the presentations in these proceedings have been summarized either by the authors or the editors or both.

A.M. Herzberg
I. Krupka
March, 1998

Acknowledgements

The conference would not have reached fruition without the advice and help of many. Mr. Alan Farant, Front Line Television and the Acton Society Trust, U.K., provided encouragement and ideas which helped to complete the program. Professors G.A. Barnard, Sir David Cox, H.E. Daniels, D.B. McLay, J. Meisel and A.T. Stewart; Kingston's Member of Parliament, P. Milliken; and Dr. G. Herzberg were involved from the beginning. Their advice has been invaluable.

Through the kind offices of the Vice-Principal for Research at Queen's University, George Hood provided enthusiasm and a great deal of support during the many levels of preparation.

Thanks are due to the Berry Brothers and Rudd, Ltd., The Davies Charitable Foundation, Falconbridge Limited, the U.K. Foreign and Commonwealth Office, Vasari Enterprises Limited and many anonymous donors for their financial support.

Jean Whittle, Sally Adams and Jane Adams were instrumental in the organization both at Queen's University and at Herstmonceux Castle. Marilyn Banting and Shirley Fraser helped with the editorial work and Peter Dorn did the graphics.

Contributors

B.A. Bailar, University of Chicago
J.C. Bailar, III, University of Chicago
D.J. Bartholomew, London School of Economics
D.V. Bates, University of British Columbia
T.A. Brzustowski, Natural Sciences and Engineering Research Council of Canada
P. Calamai, *Ottawa Citizen*
L.L. Conquest, University of Washington
Sir David Cox, Nuffield College, Oxford
P.B. Dixon, Monash University
S. Fortier, Queen's University
F.D. Gault, Statistics Canada
E.A. Gehan, Georgetown University Medical Center
J.F. Gentleman, Statistics Canada
D.J. Hand, The Open University, U.K.
A.M. Herzberg, Queen's University
C.C. Heyde, Columbia University, New York, and Australian National University, Canberrra
Sir John Kingman, University of Bristol
I. Krupka, Pest Management Regulatory Agency, Canada
J.F. Lawless, University of Waterloo
W. Leiss, Queen's University
D. McDonald, Monash University
J. Meisel, Queen's University
P. Milliken, Member of the Canadian Parliament for Kingston and the Islands
G.H. Reynolds, Centers for Disease Control and Prevention, Atlanta
A.T. Stewart, Queen's University
R.E. Taylor, Stanford Linear Accelerator Center

PART I

INTRODUCTION

CHAPTER 1

Opening Remarks

A.M. Herzberg

In late 1992, Queen's University in Kingston, Ontario, Canada was given Herstmonceux Castle by Dr. and Mrs. Alfred Bader. Soon afterwards, with the encouragement of many who thought it would be a good idea to have a conference at the Castle, I reserved some dates for April 1995 for a Conference on Probability and Statistics. In August 1993, I realized, as there were very many meetings on probability and statistics, that it would be better to have a change of air. The conference would be on Statistics, Science and Public Policy.

Such a conference would have thirty-five to forty-five participants. Planning and a search for funding began. Because no funds were forthcoming at first, the conference was postponed to April 1996.

I took a risk and invited individuals to come at their expense and remained hopeful, always keeping in mind the aphorism of Donald Cameron, the founder of the Banff School of Fine Arts in the Rocky Mountains: "Make no small plans; they make no magic to stir men's blood."

At the conference many current issues would be discussed including the lack of communication among statisticians and scientists, the difficulty and inaccuracy in communicating scientific results to the public through the media, the recent interest in the support of science from the public purse, and whether science should be directed by government. Many felt that the time was right and by gathering at Herstmonceux Castle, it would be possible to attract individuals from many different countries.

Scientists have in the past, and it is still true today, been underrepresented in the governments of Canada, the United States and the United Kingdom. Because of this, the voice of science must become louder.

One problem we confront is that many scientists believe that government has no place in science. They argue, for example, that the theory of electricity and the particular compound needed for the ever-present "Post-it" notes were discovered by scientists doing what interested them not what governments commissioned! In an article, Michael Smith, a Nobel Laureate at the University of British Columbia, points out that when the Nobel Laureate, Gobind Khorana, was an unknown Indian scientist in the early 1950s and was given a research grant to do basic research by the British Columbia Research Council, an applied institute, discussion arose about what Khorana should be told to work on that might be useful. Dr. Gordon Shrum, the then head of the Council, said "Gentlemen, we did not hire a man of Dr. Khorana's talents to tell him what to do" (Smith, 1995, pp. 72-73).

Because of this lack of perception and knowledge of science by elected representatives, groups of scientists in various countries have begun lobbying. In the U.K., the group "Save British Science" has been formed. In Canada, a number of protests were made against the 1990 cuts to the National Research Council. Scientists and politicians need to learn to understand and trust each other's perspective.

Scientists must keep telling politicians what the scientists think until the politicians think it was their own idea and then the scientists have to look as surprised as children with Santa Claus when something happens. Or as Edmond Burke might have said: "It is necessary only for the good man to do nothing for evil to triumph." This may be a bit strong but the point remains.

A story told by Professor D.A. Bromley of Yale University states:

> A Mexican robs a bank and escapes over the border to the United States. He is caught in a small town by a Sheriff who speaks only English and not Spanish. A teacher is found who speaks both languages. The interview goes as follows. "Ask him his name." "His name is Pedro." "Ask Pedro if he robbed the bank." "Yes, he robbed the bank." "Ask him where he put the money." Although Pedro says he put it in the town well, the translator says "Pedro will not say." The Sheriff puts a gun to Pedro's head. "Ask him again." "Pedro says he is prepared to die."

This anecdote illustrates that one should be aware of the other person's agenda; we have to learn to trust each other.

Along with the bringing of statistics and science to the public, we should also be conscious of bringing statistics and science to the attention of the government. Dr. John Polanyi of the University of Toronto recently stated,

> We, the scientists, have failed to explain the nature of the scientific and scholarly enterprises to the public, and so we have failed to communicate it to governments. I think we just haven't tried hard enough. We've despaired of making such simple points as that discovery is a high risk enterprise, and you don't go into your laboratory wondering what important discovery you're going to make today.

Thomas Hobbes once said, "There may be no contentment but in proceeding." This could be considered perhaps a benediction for this conference.

REFERENCE

Smith, M. (1995), "Synthetic DNA and Biology", in *Science and Society. The John C. Polanyi Nobel Laureate Lectures,* ed. M. Moskovits, Concord: Anansi, pp. 71-82.

CHAPTER 2

Science, Statistics and Public Policy

Sir John Kingman

The Royal Greenwich Observatory (RGO) was, of course, originally at Greenwich, on the hill above the beautiful palace that later became the Royal Naval College. It was run by the Admiralty because it had been established by King Charles II to solve the longitude problem, to give the captains of His Majesty's Ships an accurate idea of their positions on the high seas. As time went by it extended its activities into much broader areas of astronomy, under the leadership of a distinguished series of Astronomers Royal. It nevertheless retained its expertise in the accurate measurement of time, which is why international agreement on a global time framework came to be based on the Greenwich meridian, the line of longitude passing through the centre of one of the Greenwich telescopes.

Greenwich became less and less suitable for serious astronomy as the lights of London drew nearer, and after the Second World War the Admiralty looked for another site, preferably pleasantly rural, cheap, and near the meridian. They found Herstmonceux Castle, and re-established the Royal Greenwich Observatory here. It was not a very clever decision: the climate is too cloudy and the lights of Eastbourne are too close for modern telescopes to be effective. But the Astronomer Royal reigned here in great state, and emitted each hour the Greenwich Time Signal familiar to all listeners to the BBC.

All good things come to an end, and in 1965 there came about one of those reorganizations of scientific research which governments use as a substitute for serious strategic thought. A Science Research Council (SRC) was set up and several research establishments, including the RGO, were transferred to it. The SRC also inherited another Royal Observatory in Edinburgh, and set about developing a policy for providing the astronomical community with telescopes that could advance the subject.

It was rapidly realized that there was no point in building modern telescopes in the murky climate of Sussex, or even Edinburgh, and the Council built new instruments high on mountains in clearer parts of the world, in Australia, Hawaii and the Canaries. The Castle became the homebase for a splendid complex of telescopes on La Palma in the Canaries, which has proved excellent value as a rich source of astronomical data.

I tell this story by way of introduction to the themes of this conference. It is about getting value for money in scientific research. It is about the way a particular applied problem can open the door to a new field of basic research. Or to put it more brutally, it is about the way in which scientists can divert resources intended to solve practical problems into useless speculation about the nature of the universe. It is about international collaboration in science — the La Palma observatory is a seven-nation project — and about the economic consequences — who got the building contracts?

Just as politicians take decisions for reasons which have little to do with real advantages, so scientists display curious priorities when faced with difficult decisions. When the telescopes were built away from Herstmonceux and Edinburgh, the rationale of these two establishments became very doubtful, and there seemed every advantage in a single U.K. base for all the overseas observatories. Where should it be? Not here at the Castle, isolated in the Sussex marshes miles from the nearest university. Edinburgh was the obvious answer, but English astronomers could not bear the thought of being ruled from Scotland. And so the Royal Greenwich Observatory moved to Cambridge!

All advanced countries are agonizing about issues of this sort: How much to spend on basic science and in which areas? How to ensure that the advances of science can be put to practical use? The extent to which international collaboration is useful or inevitable. How to get the best value for money? In general, governments want to spend less and scientists more. The idea that the best scientists should be supported because their instincts are the soundest guide to where the new discoveries are to be found (the truffle-hound principle) is regarded by treasuries as the crudest self-indulgence. Our rulers want to know in advance where science will lead, and are not pleased to be told that science does not work like that.

These questions have been the subject of an excellent U.K. government report, a White Paper entitled *Realising our Potential.* It starts from the basis that science should contribute to wealth creation and the quality of life, admits that these are long-term as well as short-term objectives, and insists on the need to choose between areas of science according to a careful analysis of technological challenges and opportunities. So far, so good; a country like the U.K. cannot afford to do everything, and science has never been driven by pure intellectual curiosity uncontaminated by a wish to better the human lot.

Unfortunately, the subtle analysis of the White Paper was read in the Treasury as an encouragement to pick a few winners, which could be supported while the rest of science was neglected. An elaborate Technology Foresight Exercise was mounted, with industrial and academic scientists made to vote on long lists of priorities. To the extent that it brought people of different backgrounds together to discuss the results, it was valuable, but the actual results were the least convincing output.

This should have been no surprise to anyone who conducted a simple thought experiment. Suppose that in (say) 1946, a Technology Foresight Exercise had been mounted, it would no doubt have yielded sensible thoughts about, for instance, faster aircraft, new uses for radar, improved building materials, the development of antibiotics. It would not have predicted the transistor, the microprocessor, communication by satellite, the synthesis of drugs, genetic manipulation or biotechnology. Why should we suppose that we can predict the changes to come, in the next twenty let alone the next fifty years? The best we can do is to observe that the really new contributions to wealth creation and the quality of life come out of good science, done by good scientists who have an eye for the grain of the subject. If they are not supported, we shall miss the opportunities for real progress.

These issues lie between science and public policy, but we are to discuss a triangle in which these are but two of the vertices. The third is statistics, which has links with science on the one hand and public policy on the other. It is uncontroversial that science needs statistics. It is a very rare scientist whose data are so accurate that they need no statistical analysis or validation. Some scientists resent their reliance on statistical methods, but this attitude is dying out, and many scientists are very competent statisticians themselves. Statisticians in turn find this disturbing; the barbarians invade the sacred mysteries. But it is better to have scientists who are good statisticians than that they should make mistakes that confuse their science.

The same is true about public policy. Rational decisions about a whole range of issues cannot be made unless there is an understanding of both how to get information and how to appreciate the limitations of that information. So practitioners of public policy need statistics. The problem is that they often expect the statistics to confirm their prejudices rather than to give unbiased information. In the words of the old joke, they may use statistics as the drunk uses the lamppost, for support rather than illumination.

It has to be said, however, that there are strong national differences here, and Canada stands out as a model of good practice. As I understand the situation, the independence of Statistics Canada is enshrined in statute and buttressed by a National Statistics Council of outside advisers. By contrast, the British constitution defines the national interest as the interest of the government of the day, and gives each public department a function towards which all its activities must be directed. What is remarkable is that the British statistical service has maintained such high standards when its legal status should subordinate objectivity to the requirements of day-to-day government. It cannot be taken for granted that these standards will continue without constant vigilance.

These are some of the issues that lie, as it were, along the edges of the triangle. There are some general points which I should like to shade into the interior, because they draw together all three of the topics of the symposium. One is the absolute need for professionalism, for responsibility, for an honest facing of difficulties rather than the quick (and popular) fix. Statisticians argue about whether there is, or should be, a statistical profession. Not in the sense that the doctors or the accountants form a profession, a monopoly which restricts certain activities (and the resulting profits) to their members. You do not need to be a Chartered Statistician to carry out an analysis of variance. But when a statistician makes a statistical report, the statistician needs to be trusted to have used the best possible methods and experience to give the best possible analysis of the data. This trust must be built up and protected, as a collective asset of the whole statistical community.

Statistics is not just a matter of applying a few standard formulae, or a piece of standard software. The subject took a very wrong turning when it started to believe that there were mathematical theorems which laid down the ways in which inferences should be made. A statistician needs to ask where the data come from, whether they have structure which can be exploited, what sort of random corruption is likely, whether there are plausible underlying models, before attempting to arrive at reliable conclusions. These are not mechanical procedures, even though there are general principles and good practices to guide one. A mature and experienced judgement must be present, along with impartiality. If a statistician abuses his position, he will not be trusted, and the same is true of the scientist or the policy maker. In the short run, this may be an acceptable risk, especially for the politician whose horizons are limited by the next election. And the temptations exist in other fields; the scientist must gain grants for research, and the statistician may feel the need to compromise to achieve a position of influence. But a short-term victory will be paid for in loss of integrity and trust: "Lies, damned lies and Statistics!"

. . . .

Following the presentation of Sir John Kingman's paper, the following significant points were made during the discussion.

Mr. D. Wroe questioned the speaker's interpretation of the constitutional position of British government statisticians. He explained that, like all civil servants, they were bound to offer impartial advice to ministers. Sir John Kingman replied that he believed that this advice was expected to be based on government policy, and that there were pressures for statistical advice to justify, rather than to undermine, that policy. They agreed that in practice, government statisticians were able to preserve their integrity, but in Sir John's view, this was despite, not because of, the legal position.

Professor J.M. Ziman stressed the importance of basic research and of independent scientific advice to government of the sort that could come from good university laboratories. Sir John agreed and pointed out that, when a new practical problem arose, there was no time to grow experts; they had to come from existing research groups.

Dr. D.V. Bates praised the American system for the openness of advice to government, and cited his experience in 1968 before the U.S. Senate committee chaired by Senator Muskie. Dr. J.C. Bailar drew attention to gaps in that system. Every system has strengths and weaknesses and none is perfect.

Professor D.J. Hand questioned whether scientists made good judgements of research priorities, or whether peer review led to specialization into small and irrelevant problems. Sir John Kingman claimed that peer review bodies usually rejected such projects and that the danger might be in the other direction, that they might, because of conservative bias, miss the unusual new ideas. Research support must be regarded as risk money.

Professor E.B. Anderson, after describing the situation in Denmark, questioned whether any system had much chance of providing good advice.

Dr. J.F. Gentleman introduced a lively discussion about the incorruptibility or otherwise of university scientists who had to find sources of financial support for research, an issue also addressed by Professors J. Meisel and A.F.M. Smith. It was argued that, while there might be temptations for individuals, it was quite difficult for scientific corruption to remain undetected.

Professor J.F. Lawless asked about the attractiveness of research careers to young men and women today. In reply to comments by Drs. T.A. Brzustowski and F.D. Gault, Sir John argued that science policy should be based on a pyramid of ideas and developments. At the base it should encourage a wide variety, but only a fraction of the new ideas should go into development, where costs are so much higher. Of these, even fewer will enter the market, but this is not a reason for premature selectivity.

PART II

COVERAGE BY THE MEDIA

CHAPTER 3

Why Journalists Can't Add

P. Calamai

INTRODUCTION

Journalists get science wrong far too often in the media, and statistics wrong even more. On the little things, it is mostly our fault. On the big things, however, it is mostly the fault of scientists, public officials, politicians — and even statisticians.

What are the big things? They are issues such as silicone breast implants, the downsizing of the middle class, gender pay equity, the consumer price index, violence against women, the poverty line, drinking and pregnant women, race and crime, international debt comparisons, unemployment rates, the disappearance of the Atlantic cod, the tainted blood scandal and, of course, BSE and CJD,[1] two sets of initials known only to a few researchers until Mad Cow Disease hit the headlines in 1996.

What are the little things? They are absurd metric conversions, meaningless decimal places, constant confusion between incidence and prevalence, the shoddy presentation of all public opinion survey data, repeated misuse of per cent for percentage point, an utter ignorance of conditional probability, silly horoscopes, daily embracing of Simpson's paradox and so on.

In order to understand why journalists think only the little stuff is our fault, non-journalists need to know the two cardinal media truisms: first, the urgent always drives out the important; and secondly, journalists consider their job largely done when they find a credible source. In practice, these two truisms mean that if someone associated with a university produces a statistical study of an urgent nature, it will immediately get ink and air time, no matter how obviously flawed to anyone knowledgeable in the field. For example, consider the 1993 study of campus date abuse that concluded that four out of five women at Canadian universities had been abused by men they dated in the past year. That became front-page news.

However, the largest single contributing factor to the four-in-five statistic turned out to be the three-in-five women who answered Yes to the category "Did something or said something to spite you." The second largest contribution came from "insults or swearing" (half the women said Yes) and the third was "accused you of having affairs or flirting with other men" (almost two out of five women said Yes). All these categories are components of the Conflict Tactics Scale, a methodology whose shortcomings had been documented by 1993 in more than twenty research papers and which was rejected as unsuitable by Statistics Canada for its own Violence Against Women Survey. The result of using this unreliable methodology was a news story that exaggerated a real problem so much it instead wound up being trivialized. Whose fault? The researchers who should have advertized the shortcomings of their methodology or the harried reporter who had not read twenty research reports on the Conflict Tactics Scale?

The problem is, despite what non-journalists may have learned at the movies, most journalists do not consider their job as "getting the truth"; they are satisfied if they can just get a few facts. And what better facts than statistics from a supposedly credible source. After all, our Western democratic societies are structured to take action only on problems that can be measured. Some have called this attitude a tyranny of numbers. All too often it is a tyranny of bad numbers produced by people with particular social and economic agendas, people with axes to grind.

There are many examples of shaky statistics combining with bad politics. Take the consistent under-reporting of jobless rates by ignoring those so discouraged that they stop looking for work. The story of how Canada's fish researchers managed to miss the disappearance of the Atlantic cod is worth an article in itself. And then there is the poverty line that measures the wrong thing; household income, rather than getting at consumption through purchasing power, expenditures and non-cash benefits.

Issues like these are battles in a high-intensity public policy war, with the combatants constantly being mortared by statistics. The problem is that journalists are essentially unarmed combatants, since they are largely innumerate. In addition to such external factors, there are three internal reasons why the media all too often get it wrong: they *don't know how*; *don't want to know*; and *don't challenge*!

JOURNALISTS WHO DON'T KNOW HOW

People who become journalists are self-selected. Their positive characteristics include: curiosity, love of words, desire to tell a story, and an inchoate commitment to public betterment. Their negative characteristics include: short attention span (ephemeral, in the true sense) and a fear of numbers.

A few years back, Michael Cobden, who was directing the journalism program at King's College in Halifax, asked the entering students why they had selected journalism. The most prevalent answer was to get a job, followed by (i) because I'm no good at mathematics and (ii) because I'm no good in science.

Not surprisingly, most journalists are intuitive rather than analytical. Newsrooms at papers or broadcasting stations are populated with people who breathed a sigh of relief when they scraped through their last course in mathematics and who avoided all sorts of other careers because they had to understand mathematics. If you whisper the word fraction, most journalists become catatonic. Assume, as a given, that almost no one who works in a newsroom today, certainly no one under forty-five, can work out a percentage or understand metric conversions. The dismal results are on view daily. Nor is this problem getting much attention at journalism schools because journalism professors also tend to be intuitive rather than analytical. Little wonder that science writing rated third from the bottom when Canadian journalism schools were surveyed in 1992 about speciality reporting courses. The only ones lower were environment and technology; see Table 1.

There are some solutions to the "don't-know problem". Simple statistical "cheat sheets" in newsrooms would stomp out many of the silly mistakes, like confusing the consumer price index (CPI) with the Bank of Canada's "cold and inedible" CPI-LFE, that is, less food and energy.

And every newsroom library should have three volumes: *News and Numbers* by V. Cohn, *A Mathematician Reads the Newspaper* by J.A. Paulos, and *Precision Journalism* by P. Meyer. Finally, granting bodies might want to look at two models that could be replicated elsewhere: (i) the Council for the Advancement of Science Writing in the

Table 1: Canadian Journalism School Courses (1992)

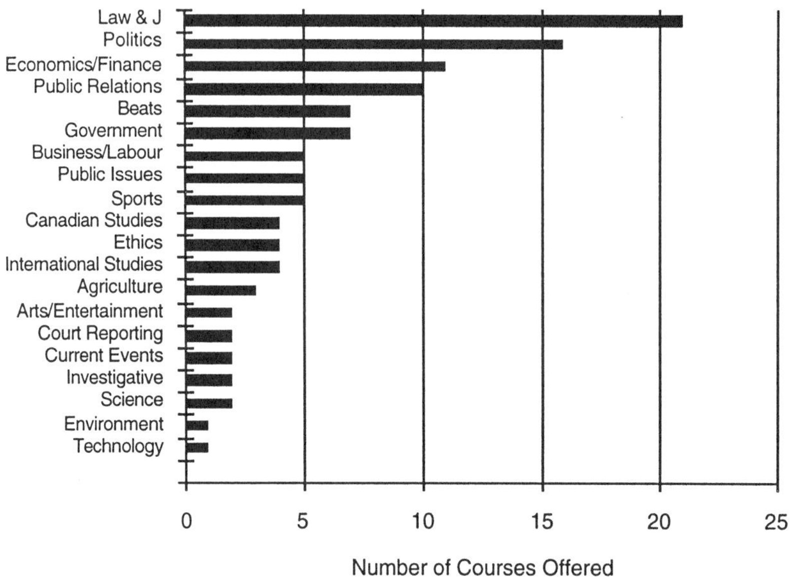

Source: The Impact Group

United States, which for a long period was funded by the National Science Foundation; (ii) the briefing sessions organized for journalists on environmental issues in the 1970s and 1980s by Earthscan, an offshoot of Barbara Ward's International Institute for Environment and Development.

Even with all that, a far greater problem remains.

JOURNALISTS WHO DON'T WANT TO KNOW

These journalists fall into three subcategories: (i) journalists who will ruin a perfectly good story; (ii) journalists with ideological bias; and (iii) journalists with market considerations.

Ruin a Good Story. Fear of numeracy ruining a perfectly good story is seen most often in the treatment of public opinion surveys. All but a few newspapers and broadcasters steadfastly refuse to admit to readers that the greatest sources of error in such surveys lies in the wording of the questions, the drawing of the sample, the treatment of non-responses and the recording of responses. Instead, all too often readers and viewers are deliberately misled by statements that the findings will be statistically valid, plus or minus four percentage points, nineteen times out of twenty. Using these confidence levels assumes that everything else has been done perfectly, which is highly unlikely.

Worse, some journalists, not knowing any better, pile further innumeracy atop such deception. *The Ottawa Citizen* once carried a main front-page story based on a mail-out survey of doctors, to which fewer than half responded. The story carried a standard sentence saying it "would be considered accurate within three percentage points, nineteen times out of twenty." Nonsense, of course.

Polls are not the only type of story where journalists do not want the facts to get in the way. Take a front-page story in *The Toronto Star* on 27 July 1995 under a headline proclaiming that "Thousands of students shun universities." The first hint of deliberate blindness is the absence of actual numbers, for applicants to universities the previous and current years and, the more important number, the total eligible to apply. Instead, the *Star* story throws around percentages, "universities are experiencing a 5.1 per cent average drop in the number of applicants for first-year, full-time spots". As I suspected, two percentage points of that drop can be attributed to demographic factors that had been widely forecast, the declining age cohort and participation rate, and not to students "shunning" universities. And more accurate counting reduced the gap further. The real drop was no more than 2.5 per cent, which worked out to 1,468 students fewer than if the same proportion of those eligible had applied in 1995 as in 1994. So much for the thousands in the headline and story. I informed the *Star*'s Bureau of Accuracy. No correction appeared.

Ideological Bias. At least the innumeracy in reporting polls is not usually ideologically driven. I wish I could say the same for other reporting of dubious statistics.

I have already described the seriously-flawed study of campus date abuse, which was uncritically accepted. First, because it came from a credible source. And secondly, in my view, because many journalists wanted to believe it to be true, so they relaxed their critical reporting facilities.

Even more ideologically suspect are the constant articles in North America media about the male-female wage gap, that always report that women earn only 69 or 70 or 71 cents for every dollar men earn. Yet both the International Labour Organization and U.S. Department of Commerce demonstrated as far back as 1992 that this bald statistic conceals more than it reveals and distorts more than it informs. When you compare women in the same jobs with men, with similar education and years of experience, the wage gap shrinks to as low as five cents an hour. This has been admitted by the Public Service Alliance of Canada, even though this large public federal union is engaged in a drawn-out wage equity dispute over a claimed $1.5 billion in back pay.

The real story here is the pink ceiling: women do not get to those high-paying jobs, they do not get the education, and they do not rack up the consecutive years on the job because of child-bearing gaps. These are the real issues that need to be addressed. Instead, insupportable and repeated news reports claiming that women are paid 30 per cent less than comparable men in the same jobs divert the public's attention.

And then there is Naomi Wolf, author of *The Beauty Myth*, who claimed 150,000 women were dying of anorexia a year in America, proof of her thesis that women are so subjugated by society that they can starve to death. That statistic wound up in countless news reports, columns and even editorials. Then Christina Sommers, an associate professor at Clark University in Massachusetts decided to check out the figure. Her book *Who Stole Feminism: How Women Have Betrayed Women* reports, with citations, that the actual number of deaths is fewer than one hundred. Guess which figure is still being regularly cited by women journalists?

Ideological bias works the other way as well, in helping decide what statistics are not collected. Because the Metro Toronto Police Board refused to let police report the race of persons charged with violent crimes, many Metro residents think they are in imminent danger of being gunned down by drug-crazed blacks as they eat their dessert in a restaurant. And when people with agendas cite their own estimates of the crime rate, reporters have nowhere to turn for more reliable and less-biased data.

Market Considerations. Of course, we all know that astrology and horoscopes are scientific nonsense. As John Allen Paulos writes, "Their aptness is self-provided. In effect, their cryptic obscurity provides a random set of 'answers' that the devotee fabricates into something seemingly appropriate and useful."

In December 1995, the Sunday magazine of the London *Independent* printed a major feature pointing out how useless newspaper horoscopes are for anyone who actually believes in the theory. A few days later the daily *Independent* newspaper printed two full pages of horoscopes for the next year. At the top of every astrology column, *The Ottawa Citizen* inserts a disclaimer that the column is published "solely for the entertainment of readers". Newspapers and magazines know horoscopes are scientific nonsense, yet we publish them. Why? Because they are immensely popular and profitable. With newspapers adding pay-per-call audiotext services, astrology columns are now subsidizing the salaries of science writers.

JOURNALISTS WHO DON'T CHALLENGE

Sometimes it is not that journalists cannot add or do not want to know for one reason or another. Sometimes we are misled by people who should know better and we do not challenge the statistics that we should.

A trivial example is the best-seller book list that appears in newspapers in Canada, Britain and the United States. We really should not print them, because they are terribly misleading. In Canada, for example, they exclude trade paperbacks; in the U.S., sales by the Book of the Month Club are not counted; the British list omitted the phenomenal sales of Penguin's mini-books.

Less trivial is the fudging that has been taking place since last year with the consumer price index in Canada. Statistics Canada still issues CPI figures for individual cities, even though only the shelter costs are specific to the cities (cutbacks mean only sixteen are surveyed rather than the initial eighty-two). Other components such as food and transportation costs use provincial averages. So why continue to produce city indexes known to be of a lower quality? Because many union contracts and spousal support agreements have escalation clauses based on the city CPI produced by Statistics Canada. This is a sham and journalists should be saying so.

And decidedly untrivial is Britain's continuing example of the BSE-CJD link. This has been the classic cock-up of statistics, science and public policy. First, there were officials from the Ministry of Agriculture,

Fisheries and Food rubbishing people like Professor Richard Lacey through the safety of not-for-attribution briefings for journalists. For years Lacey thought BSE could jump species to infect humans. He turned out, tragically, to be right. Then there was the deliberate suppression of the research finding in 1992 that BSE had passed to humans. The neuropathologist at Newcastle General Hospital who made the finding, Dr. Robert Perry, was labelled "a loose cannon who could cause mayhem" in memos from officials of the Public Health Laboratory Service. Then finally there were ministers of the Crown treating the public, in one British editorialist's words, as "ignorant, stupid and innumerate".

Unfortunately, in the BSE-CJD incident, most journalists (with several honourable exceptions) did not challenge the government version early enough and did not ask why scientists alone were allowed to deliberate behind doors a matter where public confidence was as important as public understanding. Instead, we largely fulfilled our traditional role of waiting until the battle was over and then riding down to the field and shooting the wounded.

CONCLUSION

Obviously, it is important to get statistics and science correct. Just as obviously, journalists have a pivotal role to play.

But it is not only journalists who mess it up. Here are two more cases where the problem was created by people who should have known better, scientists and public officials.

Carleton University in Ottawa became known across the country as "Last-Chance U" because a senior administrator botched the statistics needed for a ranking of universities sent to *Maclean's*, Canada's national news magazine. And tens of thousands of Canadians got let off pending criminal charges in the early 1990s because the Supreme Court of Canada got its sums wrong in the *Askov* decision about undue delay in trials.

Amidst all this turmoil and finger-pointing, journalists have one problem that is none of their direct making. They cannot assume their audience has even the most basic background knowledge about science. In fact, reasonably rigorous surveys of the scientific literacy of 2,000 adults in each of Britain, the United States and Canada in 1988 and 1989 produced fairly depressing results (Table 2). Only 31 per cent of Britons surveyed, for example, correctly answered that electrons are smaller

Table 2: Canada, United Kingdom and United States: Comparisons on Knowledge of Basic Science Ideas

	Percentage Who Responded Correctly		
	Canada	*U.K.*	*U.S.*
The centre of the earth is very hot	84.4	86.3	80.3
The oxygen we breathe comes from plants	80.4	59.9	80.6
Radioactive milk can be made safe by boiling it	61.3	65.1	64.1
Lasers work by focusing sound waves	38.0	41.8	36.0
Sunlight can cause skin cancer	95.5	93.5	96.9
Hot air rises	96.0	96.7	97.0
Human beings as we know them today developed from earlier groups of animals	58.0	79.0	51.7
Electrons are smaller than atoms	46.7	30.9	42.7
The earliest humans lived at the same time as dinosaurs	45.9	46.2	36.8
The continents are moving slowly about on the surface of the Earth	74.9	71.7	80.1
Which travels faster, sound or light? Light travels faster	73.8	74.7	76.1
Does the sun go around the earth or does the earth go around the sun? Earth around sun	78.4	62.8	72.5

Source: Einsiedel (1990, p. 14, Table 7).

than atoms, only 38 per cent of Canadians knew that lasers work by focusing light waves and only 37 per cent of Americans were aware that the age of dinosaurs ended long before the emergence of humans.

What this all means is that restaurant critics can assume their readers have more basic background knowledge than can science journalists.

So with education doing such an abysmal job of providing citizens with even basic science literacy, it is little wonder that fewer than half the daily newspapers in Canada reported they even had a science beat in 1992 and, in some cases, the same reporters also were responsible for medical and environmental coverage as well. I think it is asking an awful lot to expect journalists to accomplish in a story that takes ten minutes to read what a country's education system has failed to do in a dozen years or more. And it is really rich when officials from Canada's Science Ministry complain about scientific illiteracy. Their own 1996 public document on *Science and Technology for the 21st Century* pegged in at postgraduate reading level on the Flesch scale.

Certainly journalists and media corporations can do a much better job of writing about science and handling statistics. But they could use some help from the very people who are doing most of the complaining, scientists, public officials, politicians and statisticians.

NOTE

1. BSE is Bovine Spongiform Encephalopathy and CJD is Creutzfekdt-Jakob Disease.

REFERENCES

Cohn, V. (1989), *News and Numbers*, Ames: Iowa State University Press.

Einsiedel, E.F. (1990), "Scientific Literacy: A Survey of Adult Canadians", MA Thesis, Calgary: University of Calgary, Graduate Program in Communication Studies.

Meyer, P. (1979), *Precision Journalism*, Bloomington: Indiana University Press.

Paulos, J.A. (1995), *A Mathematician Reads the Newspaper*, New York: Basic Books.

CHAPTER 4

Science, Statistics and Public Policy: The Importance of Communication

J.C. Bailar, III

INTRODUCTION

If mathematics is the Queen of the sciences, statistics is surely the King. The subject matter of statistics is still widely misunderstood, but is now appreciated to focus much more on study design than on computations from results. Thus, statistical concepts and tools are vitally important in selecting and framing questions for study, choosing general research approaches to some issue, preparation of research protocols, selection of subjects for study, measuring (and then improving) the quality of data, reducing extensive observations to a usable format, and generalizing results from subjects actually studied to those in a broader universe. Along the way, the usual statistical operations of finding means, variances, confidence bounds, p-values, and other functions of the data have an important role, but that role is secondary.

This view of statistics, in which an understanding of research methods and processes is central, means that much of the statistician's tool kit is portable from problem to problem, just as mathematics is. The mathematician who can solve a differential equation that is important in (say) the design of highway bridges does just as well in applying that equation to understand a chemical process, or improve a drug for human use, for example. Similarly, the statistician who understands random sampling can draw an appropriate sample from a population of humans, hospitals, industrial firms, or widgets on an assembly line. This great generality of statistical concepts and procedures is an important source of its strength, because what is learned in one scientific field or one area of endeavour is likely to have immediate implications in many others.

This view that the statistician is the expert in the scientific method is certainly not the view of the public, nor is it understood by many of the scientists who work with the news media, or most reporters, or by most experts in public policy. Even some persons working as statisticians do not have, or are not allowed to exercise, this broader view. Thus, there is often poor cross-understanding among the three topics discussed in this volume: statistics/science, the news media and public policy. Deep mutual understanding is even more rare.

The problem of inadequate cross-communication is much deeper than lack of understanding. Some scientists, some policy makers, even some journalists want to use their interactions with other groups to promote agendas not based on any credible interpretation of what is known about the world around us. This is often done by such means as the highly selective citation of evidence, personal testimonials, or simple repetition of some opinion at high decibel levels, the "big lie".

BALANCE IN REPORTING

My primary message is that we, as statisticians and other scientists, have two different and in some ways contradictory problems in communication with policy makers. First, we overstate the importance and certainty of our findings taken one by one; secondly, we do not sufficiently stress the importance of looking at the whole picture. Thus, each new and intriguing bit of evidence may be seriously overinterpreted. As a sequence of studies converges to the right scientific answer by successive approximations, the public has an impression of wild swings in the evidence, with each new report negating all that came before. The news media and policy makers are willing accomplices in both of these problems, but generally they are no more than accomplices. In almost any aspect of science (as in daily life), we never know as much as we think we do. We, the statisticians and scientists, are doing ourselves in by overstating our most recent result and understating how it adds to a structure of evidence already present.

Good reporting requires both accuracy and balance, and balance requires not only weighing the strengths of opposing views and reporting them; it requires reporting of these weights. Evenly matched sound bites from both sides (or all sides) do not do the trick. Further, as the evidence converges, it will at some point be sufficiently convincing that the views presented by nay-sayers lose their credibility and no longer merit any mention at all. An example of this is the ritual quoting of tobacco interests in every news story about health hazards that have already cost millions of lives and have been solidly established for decades. An overarching emphasis on balance may damage efforts to create an informed society. It has surely contributed to some destructive public debates (health hazards from silicone implants or electric power lines; cellular telephones and brain cancer), and to the spread of ideas that are actively anti-scientific (creationism, astrology).

Many newspapers carry a daily astrology column for no better reason than that some readers want to took at their horoscopes. Some observers who know that astrology is bogus science may regard horoscopes as harmless; I do not. Teaching readers, by example, to believe that there is some merit in astrology can undermine the real scientific understanding of a public that is ever more dependent on science. It will be hard to erase memories of astrologic advice affecting decisions in the White House, and editorial judgements about horoscopes surely contributed to the atmosphere that led to that embarrassing spectacle.

There have been numerous news stories recently about possible hazards of electric power lines and other sources of electromagnetic radiation, with special concern about leukemia and other forms of cancer in children. It is not so widely reported that every expert group that has looked at this matter has concluded that, first, it is physically impossible for the energy levels involved to have had any biologic impact, and secondly, the data do not suggest an effect anyway. Billions of dollars may now be spent to protect the public from a non-existent hazard.

Breast implants have been widely suspected of causing health problems; it is interesting that in the United States concern has focused on the silicon content of the implant, whereas in Canada it has focused on the coating. There is no credible evidence to support either of these concerns, though the public, led by lawyers who profit from the matter, continues to file suits.

As a result of animal experiments, cyclamates (a class of artificial sweeteners) were banned in the United States. At about the same time, and based on similar evidence, saccharin was banned in Canada. The only reason I have found for this odd pair of regulations is in the timing: neither country wanted to ban artificial sweeteners entirely, so after the first was banned (cyclamates in the U.S., saccharin in Canada) the other was allowed to remain on the market.

The health risks of nuclear power are very widely misunderstood and overrated, and again, "balanced" news reports have surely contributed to the problem. This situation may differ from the previous examples in that claims of great hazard are being used as a form of protest against something quite different, that is, against the whole of industrial technology in the modern world. Protesters can scarcely complain effectively about the manufacture of automobiles, or tin cans, or computers, or about the availability of more abundant and varied food, or improved housing technology. But they can protest against that nuclear power plant back in the hills as a surrogate for these and many other things.

Radiation sterilization of foods elicits a strong reaction because of the word "radiation", and as a result healthful and inexpensive preserved food is denied to the public in many countries, despite the absence of any evidence whatsoever that there can be a hazard. The public seems not to understand that the radiation comes and goes during processing, and that absolutely no residual radiation remains in the processed food. Somewhat more sophisticated concerns about radiation-induced chemical changes also seem to lack any credible scientific support. Public understanding of the evidence for and against the risk of radiation of foods seems to be close to nil.

I am not denying the sincerity of persons who hold these and various other non-scientific views. It is in fact their sincerity that makes them so dangerous. They are hard to discourage; they are often resistant to information that is contrary to their beliefs; they have the drive and persistence of those who want to save the world. The news media do a grave disservice in continuing to promote such ideas once they have been effectively refuted by thorough, unbiased, independent reviews of adequate data.

NEWS MEDIA AS A WEAPON

Unfortunately, many scientists and statisticians, and even more some policy makers, want to use the news media as a weapon, a tool to reach non-scientific (often policy) objectives. Conversely, and despite their professional responsibilities, some journalists become advocates and try to use science for their own policy objectives. An American columnist, Colman McCarthy, used news reports about bovine spongioform encephalitis (BSE) as a starting point to promote his views on vegetarianism, and health fads generally thrive on the highly selective citation of bits of science.

What matters most here is sound public policy, but what is that, and how can we recognize it? The development and implementation of good public policy in a democratic society requires, among other things, the availability of a response that is politically feasible (that is, acceptable to the public); technically feasible; adequately debated before the public, with enough information of adequate quality, freely available; cost-effective; adjustable as circumstances change; and leaving an avenue for rapid withdrawal in case of serious error (such as unexpected side effects).

As a step towards developing more appropriate roles of statistics in science and public policy, greater attention to several matters would help scientists to understand the potential contributions of statistics in: (i) assuring that questions are asked in ways that can be answered efficiently, and with a minimum of uncertainty; (ii) choosing experimental approaches and designs that are not only technically sound but easy enough to understand for non-experts to grasp the implications of findings; (iii) drawing up protocols, sampling, monitoring, and improving the quality of data; (iv) converting, reducing, and presenting data; (v) statistical inference (p-values, confidence bounds); and (vi) generalizing results.

Underlying each of these items is the notion of bias, that is, noxious influences on the data such that the findings do not mean what we think they mean, or what we want them to mean. In my experience, bias is almost always a much greater threat to the validity of statistical conclusions than randomness, though my colleagues in statistics spend nearly all of their time and effort dealing with randomness.

CONCLUSION

Finally, I would like to list some steps that might help. For *statisticians and scientists*, these include not overstating the importance or the certainty of any single research finding, especially one's own. Any research project is just a single small brick in a very large wall, and cannot be understood or appreciated without consideration of the wall as a whole, with special attention to any well-made bricks that do not seem to fit. Remember, we never know as much as we think we do. Stress the need for reporters and the public to look at the whole of the evidence, and at how new information affects the whole. Promote the training and experience of scientists and statisticians in communicating with the news media and policy makers. Educated practice may not make perfect, but it can certainly help.

For *reporters*, stress the accuracy of overall impressions; do not misinform the public in the name of "balance". The defining characteristics of a lie, and what makes it reprehensible, is deliberate deception, and deliberate deception does not require misstatement of fact. One can lie quite effectively by means of artful omissions, skewed emphasis, inadequate attention to making communication understandable, and in many other ways. These are still lies. Mistrust and avoid anecdotes: anecdotes rarely tell much about a matter of great importance and may be seriously misleading; public debate by sound bite is an oxymoron.

For *policy makers*, do not misuse science for the purpose of promoting policy that is considered desirable for other reasons; what is good for small steps forward may be very bad for big steps forward. For example, incremental improvement of flawed systems may seem appropriate, but result in the end in a system that is still flawed. The problem is compounded if an appearance of incremental progress relieves legitimate demands for more fundamental change.

I have heard and sometimes participated in other discussions about the relation between science/statistics and the news media and between science/statistics and public policy. I assume that there have been similar discussions about the relation between the news media and public policy. However, this is the first occasion that I know about in which the three have been brought together. There are many rough edges in our understanding of and communication with each other. We do not even have common sets of terms, principles, concepts, or ground rules. Thus, we are beginning with a mass of detail from which we must try to make some sense. I hope and expect that there will be future similar conferences, where there may be an appearance of greater progress, but that progress will depend critically on our meeting here and on other meetings that will surely follow.

CHAPTER 5

The Chasm Between the Scientist and the Media

C.C. Heyde

INTRODUCTION

My personal experiences with the media have not been especially reassuring. Each time, the story has been misinterpreted in some way or other. Not that I should claim to be blameless in this: the scientist rarely has a straightforward story to tell.

Certainly there are significant differences from country to country, but in my experience there is a general view among decision makers (parliamentarians, civil servants, leaders of industry, etc.), and media agenda setters, that scientists have not told their story convincingly enough to warrant higher priority in the competition for public attention and support. On the other hand, among the scientific community there is the widespread belief that the nature and achievement of its work, and its potential for supporting national development, are not well-enough understood by the community at large.

There is much truth in all of this and it is worth looking at some of the principal causes.

SCIENTIFIC AND TECHNOLOGICAL KNOWLEDGE

It is nearly forty years since C.P. Snow roused his readers with a persuasive attack on the cultural chasm dividing the worlds of the humanist and the scientist. But in the intervening years the rift has probably widened. In the United States, for example, only some one per cent of the population have completed degrees in scientific and technological fields. Yet most public policy issues involve judgements that depend on knowledge from those fields. Countries have the challenge of preparing citizens who are knowledgeable in both worlds. What we have in practice is significant scientific illiteracy among humanists and tunnel vision among scientists.

Education in schools is a critical component. In Australia, school runs over twelve years: mostly six in primary school and six in secondary. A minority of Australian primary schools have whole-school science programs but in secondary school, science and mathematics are typically compulsory to the end of year ten (around age sixteen). There is quite a substantial broadly based science education in years seven to ten. Social science and humanities subjects, history, English, etc. are also compulsory to year ten. These arrangements seem to me to be minimum requirements. I think we have to be very wary of narrow specialization.

By and large journalists have a humanities background and many papers have no specialist science writer. But most scientists are little concerned with communicating to the general public, only with communication to other scientists. This means that by default the job falls to journalists who are often ill-equipped to carry it out. I do not know if this problem is universal. But I have seen it reported that the former Soviet Union published approximately six times as many popular science magazines as the United States, and most leading Soviet scientists wrote for them; it was considered part of their job. Is it coincidental that the Russian language lacks a word for scientist; I am told that one says "uchyon", literally "scholar".

Of course there are excellent popularizers of science in the West, but they tend to receive rather little credit from their peers; all too often the response is negative for time spent away from research. We should really value the good communicators and popularizers among us. Scientists must not abandon this responsibility to journalists.

REPORTING ON SCIENCE

The most popular scientific topics in the media are undoubtedly medicine and astronomy. Health is an emotive issue to people and topics such as cancer risk, fats and cholesterol are widely followed. Astronomy catches the public imagination. Recently we have had the discovery of planets moving around suns elsewhere in the galaxy (one thirty-five light-years from earth) and alternative life forms elsewhere in the universe are beginning to look entirely feasible. Biotechnology and even particle physics are popular in the media with the discovery of the top quark and generally the search for the base components of matter.

Newspaper articles on scientific subjects are skimmed by a very large number of readers. Of course, the newspapers are losing out to television so perhaps the question should be who is watching science. Science goes well on television. Nature and technical innovation programs are popular. But one difficulty is the high production costs, some $1.5 million per hour.

Science used to be regarded as an unmitigated good. Then came the Cold War and concerns over things nuclear. Also, in the early post-World War II period came a recognition of the effects of pollution and environmental degradation. Rachel Carson wrote her influential book *The Silent Spring*. Science began to be seen as a source of problems as well as solutions. Then film-makers reduced the scientist to a "long-haired and white-coated" stereotype. Nevertheless, I cannot help feeling that there is now a certain sense of public optimism that has emerged. Science will find a cure for AIDS, genetic engineering solutions to many disabilities will come, etc. Science is seen as the single-minded search for understanding, perhaps not always mindful of the social implications, but not fundamentally evil.

Some of the positive feeling in Australia has been politically induced. Former Prime Minister Hawke introduced the slogan of Australia as "the clever country". This helped to raise public confidence in science, but also ended up in raising other concerns such as lack of investment capital, foreign ownership, brain drain, etc. Unfortunately, statistics does not tend to get good press. It is towards the end of the spectrum of social perceptions as being dry, dull, boring, and often misrepresentational; politicians use statistics to mislead the public. However, opinion polls and market research have become very widespread, and the results are generally treated with appropriate caution. There now seems to be a reasonable public perception of the possible errors involved, although standard-deviation estimates are not usually published.

Fortunately many universities are already introducing joint degree programs that cut across traditional school lines. For example, there certainly are opportunities for careers in science journalism, in the print media, radio and television. At present there are limited training opportunities, at least to my knowledge. Columbia University has recently introduced a new MA/MS program that brings together the Department of Geological Sciences and the School of Journalism to train journalists in earth and environmental sciences. It would be good to see other science departments involved in such co-operative ventures, but to date only Geological Sciences has taken the initiative.

CONCLUSION

The scientific community is good at judging scientific content in research. We must broaden this to respect and reward the skill of the popularizer and communicator. And we should argue that scientists bear the social responsibility of preaching to the unconverted as well as to the converted. The community needs regular reminding that science is not largely the curiosity-driven product of socially unaware boffins.

CHAPTER 6

Improving Publicity for Science

L.L. Conquest

INTRODUCTION

An eternal lament from scientists, particularly academic scientists at a large research university like the University of Washington, is that the general public has little knowledge or appreciation for the research that we do, and views much of scientific research with great suspicion. This inevitably translates to the complaint that "we do not get nearly the research funding that we deserve".

But why should scientists even be concerned about improving publicity for science and improving public appreciation for science? First, the audience is there. People do read the science sections of newspapers, visit science centres and museums, and attend open houses sponsored by university science departments. Secondly, if legislators who control research purse strings are themselves more knowledgeable and appreciative of scientific pursuits, this can lead to a more sound public policy with respect to direction and availability of research funds. Finally, much scientific research is paid for by taxpayer dollars; we have a real obligation to let the public know how their money is being used. Expecting the public to accept unquestioning "hands-off" support for research is simply unrealistic.

HOW TO ACHIEVE ACCEPTABLE PUBLICITY

Accepting the premise that communicating research results and maintaining high visibility for scientific research is a worthwhile pursuit, how do we do it? The scientific community needs to have people in its midst who are excellent listeners and compulsive communicators.

In any scientific organization, whether a department, a public agency, a private establishment, or an academic institution, the chief executive needs to be involved in this communication. Furthermore, organization heads need to show their commitment to scientific communication by budgeting real money, i.e. more than token amounts, for news and information services.

For example, people understand scientific research endeavours more easily when they can somehow relate the research to their daily lives. At the University of Washington, in addition to the constant release of press notices to various news services, events are always taking place to communicate scientific knowledge to the public in meaningful and diverse ways. The entire Health Sciences complex at the university sponsors an annual open house as does the College of Engineering. Another highly publicized annual event is the Computer Fair held on campus. Additionally, each college within the university has its own newsletter to inform alumni, parents of students, and other interested parties of newsworthy items. Some research groups have annual symposia for targeted audiences, such as the primary funding agencies, legislators and their staffs, to inform them about current research efforts. The audience might include agency members who may actually have to recommend regulations in part based upon research results, for example, concerning environmental or public health policy. The University of Washington has also taken advantage of the Saturday afternoon schedule of (American) football games, by sponsoring "Saturday Seminars" at various places around campus. Members of the public can sign up to hear lectures on a variety of topics of interest; see Table 1. The Saturday Seminars were introduced in 1987 and have been enormously successful in terms of communicating the nature of the research enterprise to the general public. The recently published book highlighting scientific research breakthroughs accomplished by University of Washington researchers from 1882 to the present (Illman, 1996) is an example.

Regarding the broader picture, the trend concerning publicity for science is good. The print media do have science writers and weekly features about science. Certain television programs, for example, NOVA, Nature, National Geographic, are geared towards educating the public about science, as is an entire TV channel, the Discovery Channel. Radio news programs contain science features on a regular basis. And talks by scientists, properly advertised, can draw large

Table 1: Selected Topics from the Saturday Seminars Program at the University of Washington

Exploring links between media violence and societal violence
History of a sensational, unsolved murder case from 1890's Virginia
The fate and future of Hong Kong
You must remember this: repressed memory
The music of George and Ira Gershwin
Ethics of genetic engineering
The fate of Pacific Northwest and Alaska salmon environmental law in an election year
The new frontier of videoendoscopic surgery
Professional sports and the effects on the local economy
How to write a novel in only thirty years
The genetics of cancer
The dark matter of the universe
Issues surrounding the black underclass
The immigration policy debate
Winter depression and seasonal affective disorder syndrome
Comets, stardust, and planets
Effects of pollution in Puget Sound on local marine life
Underwater volcanoes, movement of the earth's crust and earthquakes

crowds, even when the speaker is not considered a celebrity. Attention-getting topics in the Pacific Northwest include anything that might relate to earthquakes (for example, volcanic, geologic, oceanographic research), fishing disputes (fisheries research, marine policy issues), human health issues (AIDS, cancer, environmental risks) and environmental and quality-of-life issues (everything from the fate of the Northern Spotted Owl to the price of timber and the cost of owning a home). Here a statistical scientist may hold something of an advantage, since one is usually working on real data, and many do research on issues that can be easily related to people's daily lives, such as public health or environmental issues.

NEW TECHNOLOGIES

One recent technological advance has made it much easier for scientific researchers to keep the public (at least the computer-literate members) informed about not only research results, but research in progress. This is the growth of the WorldWideWeb and the creation of Web "homepages" (with electronic links to other pages) for individuals and organizations. Information can be made available on the Web on a particular group's research project(s)' recent results, work in progress, future workshops, how to obtain data, the researchers themselves. The caveat here is that in order to have an attractive and informative homepage with appropriate links, the resources need to be there.

Why is it necessary to maintain public science literacy through various forms of publicity? Science changes too quickly for formal education to keep up with it. Scientists do not have to be in the teaching profession for long before realizing just how much of what is taught was not known when the scientists, themselves, were pursuing graduate and professional degrees. Therefore, informal science education is vital to keeping the public informed. And like it or not, the mass media pays a big role in popularizing science. Despite this, there are many things that scientists do not like about popular science news. Rich, complex ideas are brutally simplified to fit media constraints. Reporters get the facts or the interpretations wrong. Some members of the scientific community feel that they get punished for being very visible to the public, that once you become a "popular scientist", your credibility with your colleagues is diminished. Thus, there can be a reluctance on the part of scientists to speak with members of the press, but the result of this is that inaccurate information can appear in the news. A good public information officer can make the whole process easier for both scientists and reporters by organizing the flow of information and by handling requests for more information.

OUTREACH PROGRAMS

The University of Washington maintains an entire Office of News and Information Services, staffed by a dozen people. They are constantly looking for stories about research that will intrigue the general public. They contact different sources in various departments and read the abstracts that accompany every grant proposal that leaves the university to go to a funding agency or private foundation, The Health Sciences/Medical School complex lists a staff of nine information specialists under its director, in addition to staff for outreach events and a speakers bureau. Many research programs hire their own communications co-ordinators. An example of a major outreach project occurred when the university participated in the heavily advertised National Public Health Week in April 1996. Poster sessions, lectures and other activities for the public were held in the Health Sciences complex.

Even with a positive attitude towards journalists, academic researchers do occasionally get "burned" by ethical breaches on the part of some reporters, such as the splicing in of quotes to make it appear as though a researcher is commenting on a particular event when in fact that is not the case. Instances like these go beyond such things as misinterpretation of research results or colourful quotes that people never dreamed would end up as a headline. In such cases, complaints may be filed and a correction may be printed, but not always. Another example is that certain media outlets do not honour embargo dates on news releases, with the result that a news and information office will not send embargoed releases to those particular outlets.

CONCLUSION

Experience suggests that conscientious, active publicity for science and continuing informal education of the public is too important to be ignored. This requires dedicated resources and real co-operation from scientists. The rewards from such efforts include a better informed public and therefore better informed policy makers. I conclude with a quote from a 1995 editorial from *Science*: "Demonstrating accessibility as well as accountability to the public that pays their way and values their work is the easiest and quickest way for scientists to achieve a higher rank for science in the nation's priorities" (Woolley, 1995). Scientists owe it to the general public, and owe it to themselves.

REFERENCES

Illman, D. (1996), *Pathbreakers: A Century of Excellence in Science and Technology at the University of Washington*, Seattle, WA: University of Washington.

Woolley, M. (1995), "From Rhetoric to Reality", *Science* 269:1495.

PART III

PUBLIC POLICY

CHAPTER 7

Role of the Government in Statistics, Science and Public Policy

B.A. Bailar

INTRODUCTION

In this paper, I shall examine the role of government agencies in providing statistical survey data to the public. In many ways, government agencies shape our public policy debates. Whether it is through funding, encouragement, technology, or by providing equal, broad and timely access to data, the government is not a passive bystander. The first responsibility that government has to the formation of public policy is to provide the most unbiased data possible at the most reasonable cost, and in a timely manner so that reasonable policy debates can go forward.

Foremost among the responsibilities of the government are the following: deciding and funding data collection efforts; balancing the needs of public access to data and confidentiality provisions; setting release dates for data with strong public policy uses; providing quick and broad access to data; safeguarding the interpretation of data; and encouraging innovation.

GOVERNMENT DECISIONS ON TYPES OF DATA TO GATHER

Health

In addition to the perennial questions about smoking and health, as well as surveys that try to link lifestyles to major health risks, there are many other interesting surveys: (i) What are the uses and costs of medical and non-medical services to the HIV-positive population? (ii) How content are people who have been diagnosed with a terminal illness in their quality of life as they approach death? and (iii) What are adolescents doing to keep themselves healthy?

Education

Most of the education studies that are being funded now are large, longitudinal studies that try to assess the importance of different factors in the life experiences of young people as they move through secondary education to higher education or work experiences. Some of the questions for which we are supplying data are: (i) What are the projections of faculty in the different occupational fields in post-secondary education? (ii) What are the factors that influence children as they move from kindergarten through grade four? (iii) As people receive a baccalaureate degree and move on to graduate school, employment, or leave the workforce, what factors, courses, educational settings and family pressures influence those decisions?

Labour, Income and Finance

There are a wide range of studies in the area of labour, income and finance; many of them are longitudinal. Some of the questions being answered are: What are the issues surrounding young people's entrance into the workforce, as well as work transitions? How have patterns of wealth been changing in the current economy? How have pensions from jobs changed the economic status of families? How are families making provisions for an uncertain future?

Drugs, Alcohol and Mental Health

The areas of drugs, alcohol and mental health are becoming very active. Drugs and alcohol cost the public millions of dollars in treatment costs. Some of the studies going forward provide hypotheses for questions such as: What are the social networks of drug users? and What is known about the effectiveness of drug treatment centres?

Confidentiality versus Public Access

There is a constant struggle in all organizations carrying out government surveys between the public's right to know and the individual's guarantee of confidentiality. Confidentiality over the years has become less of a problem with written reports and text tables, as most survey organizations have become adept at controlling possible disclosures. However, data files are much more of a problem, especially with longitudinal files. It does not take many characteristics to be able to identify some very high profile individuals. Disclosure in business surveys is much more problematic than with household surveys. The government needs to ensure that no employee of a large company reading the study can pinpoint competitors.

For household and personal data, most of the disclosure avoidance has been with suppression. Some government agencies depend on random rounding. The Census Bureau experimented with "data swapping" after the 1990 census. Most often, suppression of data is used. Yet, the data element that is frequently suppressed, the location identifier, is often one in which researchers are most interested. To meet these needs some survey organizations are now making restricted access files.

A final threat to confidentiality is now coming from the ability to merge other files with a data file produced by a government agency. Suppose, for example, that the Internal Revenue Service wanted to see how individuals had reported their income to the Census Bureau. Suppose a person had a census file for which they could match individuals by means of four or five key characteristics. This might be a way to break confidentiality because the same person has a file with some of the same data elements on it. Now that there are many credit bureaus, collection agencies and other groups who have detailed computer files, this could be a problem.

Another confidentiality problem involves more groups than individuals. In the census there is a real concern that even data with no individual identifiers can reveal information that can be used for law-enforcement purposes. For example, if people reported fully, census-block reports might show that the number of persons staying in rented rooms might exceed the legal limit. City authorities could use this information as an excuse to evict. Also, many workers without legal documentation worry that full reporting may give too much information to immigration authorities.

DATES OF RELEASE

Because government agencies collect data on a wide variety of sensitive topics, it is incumbent on them to set release dates and stick to them. For many recurring surveys, such as employment and unemployment, retail trade, poverty numbers and the like, the Census Bureau has release dates and times of day set far in advance. No one can get the data ahead of time. There are at least two compelling reasons for the agreement on release dates. One is to avoid the appearance of issuing data to meet political pressures. A second reason is to give all users an equal opportunity. Many times these data are used in the stock market.

QUICK AND BROAD ACCESS

In addition to equal access, the government also needs to ensure timely and broad access to its data. Most agencies are now able to supply data in a variety of formats, from paper to CD-ROM. Users frequently call and write to agencies asking for additional data, or cross-tabulations, or combinations not previously produced. Agency staff often undertake to produce these data, often charging marginal rates. However, very little is done to keep track of these requests so that products can be revised. In fact, product development is not a strong point of government agencies.

Timely access is completely separate. Government surveys take too long, no matter if they are conducted inside an agency or by a survey organization outside the government. The government has some recurring surveys that are done very quickly. These are the monthly surveys such as the Current Population Survey and the retail trade survey, for which there are monthly release dates. However, most surveys stay in the field too long and take too long to process. Survey reports stretch out for too long. It is not unusual to release data two years after a survey begins. This is an area where the government needs to take a stronger sense of responsibility. Government agencies need to be able to guarantee their data and, if there are weaknesses, tell users what the potential problems may be.

ENCOURAGEMENT AND INNOVATION

In all areas of data provision to public policy analysts, government agencies have responsibility to provide encouragement, express the need and generally support innovation. There are now so many new technologies available to design questionnaires, collect data, transmit data and process data, that the whole survey-taking process needs to be constantly re-examined. Standard systems should be in place rather than systems tailored to each survey. These systems should be updated regularly to reflect new advances.

In some cases, government agencies are in the vanguard of the development projects. For example, the Bureau of Labor Statistics was in the forefront of having business enterprises report by touch-tone telephone. Statistics Canada pioneered standard systems for all survey activities. In most cases, however, development will occur in the private sector, but should be actively encouraged by government agencies.

CHAPTER 8

The Role of In-House Research and Analysis in Government Statistical Agencies

J.F. Gentleman

INTRODUCTION

This paper discusses the usefulness and importance of research and analysis within a statistical agency, with examples from my own experiences at Statistics Canada. It has been my observation that the word "research" in a government department title can cause the group so-named to be vulnerable to budget cuts and downsizing. Research and analysis are thought by many to be frills and are thus often among the first activities to be reduced.

Research and analysis can be defined in a rather broad sense, applying to a range of activities such as editing data, calculating and interpreting descriptive statistics, and fitting complex models. Thus, at Statistics Canada, we do research and analysis at many stages in the process of collecting data and producing information from them. The following discussion highlights the benefits of in-house research and analysis to the analysis, to the data and to the agency.

BENEFITS OF IN-HOUSE ANALYSIS TO THE ANALYSIS

The analysis is improved by being performed in-house:

(i) *In-house analysts can access data that are not available outside the agency.* In Canada, the *Statistics Act* limits appreciably the data that may be released to the public by Statistics Canada, limits that are more stringent than those in the United States, for example. Microdata are released only under special and controlled conditions and are processed to preserve confidentiality. This implies that many types of in-depth analysis that rely on the detail and rich multivariate information provided by microdata can only be performed in-house.

(ii) *In-house analysts can have earlier access to data.* Delays to the release of public-use data files are inevitable because the data must be processed and packaged beforehand. Survey data, for example, undergo editing and imputation, certification, processing to protect confidentiality, preliminary analysis and summarization, documentation, etc. before they are publicly released. In-house analysts can have access to the data before release. Indeed, they can begin to analyze the data while the collection is still underway. A successful example of this is the publication by members of my division of analyses of data from the 1994-95 National Population Health Survey almost simultaneously with the release of the public-use data file. This was possible because our analysts were given access to the data after only half of the survey interviews had been completed.

(iii) *In-house analysts can access multiple data files.* An analysis of data from one source is often enhanced by a comparative analysis of data from another source. Because Canada has a relatively centralized arrangement of its national statistical functions, centred at Statistics Canada, in-house analysts there have ready access to a large number of data sources, including the census, sample surveys and administrative data files.

(ii) *In-house analysts can link microdata files.* Linkage of two microdata files increases the richness of the multivariate information available on the separate data files, thus increasing the potential for analysis. This is such a useful activity that a considerable amount of linkage is done on a cost-recovery basis within the agency for outside clients. These activities are carefully controlled and limited so as to preserve confidentiality.

(v) *In-house analysts can have closer contact with data sources.* An in-house analyst with a question about a variable in a data file can walk down the hall to question someone in the group responsible for collecting the data. Because the experts are so handy, the in-house analyst is more likely to take the trouble to obtain the needed clarification. The census people at Statistics Canada have a particularly effective way of handling such queries, including ones from the outside; specific staff members are assigned specific census variables for which they provide expert information.

BENEFITS OF IN-HOUSE ANALYSIS TO THE DATA

In-house analysis of the data produced by a statistical agency can benefit the data themselves. Interaction between producers of data and the analysts is generally helpful in data quality assessment and data development. Close proximity of the producer and analyst of the data facilitates feedback from the latter to the former about problems with the data and about the analytic usefulness of the data.

BENEFITS OF IN-HOUSE ANALYSIS TO THE AGENCY

In-house analysis helps a national statistical agency to do its job. Most important, it helps Statistics Canada to fulfill its statutory mandate to perform research and analysis; the *Statistics Act* gives Statistics Canada the responsibility to "collect, compile, analyze, abstract and publish statistical information relating to the commercial, industrial, financial, social, economic and general activities and condition of the people".

The results of in-house analysis are widely used by the agency for policy making, priority assessment, and administrative, evaluative and scientific purposes. They are used by the agency to respond to information requests, increase the accuracy and relevance of data, increase the recognition of the agency, increase the ability of the agency to resist pressure, and increase the ability of the agency to adapt to change.

Many of these benefits were discussed by Ivan Fellegi, Statistics Canada's Chief Statistician, as part of a 1995 address to the Washington Statistical Society, a talk that was subsequently disseminated within Statistics Canada. We at Statistics Canada are fortunate to have strong support for in-house research and analysis from the Chief Statistician, as exemplified by the following excerpts from his talk (Fellegi, 1995):

> ...the authority of the Chief Statistician derives from his or her personal standing and public image. These are enhanced by speaking engagements designed to shed light on major public issues through statistical analysis....

> It is important to maintain a reasonable level of funding for methodology research, indeed to try to ensure that methodology research is integrated with methodology practice.... Methodologists are encouraged to submit research to refereed journals and to present them at professional conferences.
>
> The public profile and standing of the statistical system is an important safeguard against its politicization. A significant contributor to such standing is the *regular provision of analytic output* which reinforces a public image of relevance and helps to make the statistical office stand apart from "the government". In Canada we have placed substantial emphasis on increasing the quality and range of analytic output. We also have a policy that all data releases must be accompanied by a media-friendly analytic summary which outlines the significance of what has just been released in the context of broad economic and social developments.
>
> ... interactions [of the Chief Statistician with the media and with other deputy ministers and their departments] are far more productive if supported by a strong analytic capacity. Good analysts are needed to understand client requirements and to convey statistical findings in a manner which users can relate to their policy concerns.
>
> A strong internal analytic program contributes to an improved understanding of the needs of external analysts — in or out of government. In turn, such an understanding is a critical prerequisite of a support for new initiatives. Furthermore, good analysts have a strong personal motivation to explore issues, and can therefore champion the development of new products within the agency. They do so by drawing attention to the role that potential new information could have in illuminating an issue of public policy. In our experience this is a vital contribution to maintaining the relevance of statistical information.

CONCLUSION

In summary, in-house research and analysis at a government statistical agency is good, for the analysis, for the data and for the agency. It seems fitting to apply a quote from Bertolt Brecht (1965) to the concept of data being analyzed by those who are good for the data:

> What there is shall go to those who are good for it,
> Children to the motherly, that they prosper,
> Carts to good drivers, that they be driven well,
> The valley to the waterers, that it yield fruit.

REFERENCES

Brecht, B. (1965), *The Caucasian Chalk Circle*, London: Methuen.

Fellegi, I. (1995), "Characteristics of an Effective Statistical System", Morris Hansen Lecture, 25 October. Washington, D.C. Statistical Society.

CHAPTER 9

The View from a Research Granting Council

T.A. Brzustowski

INTRODUCTION

Statistics, science and public policy come together in many ways at the Natural Sciences and Engineering Research Council of Canada (NSERC). This paper addresses four issues: accountability for outcomes, return on investment, innovation and the evaluation process. The framework for the discussion emerges from the sequence: *inputs* into *processes* produce *outputs* which have *impacts* in the community and eventually lead to significant *outcomes*.

ACCOUNTABILITY FOR OUTCOMES

There is political pressure for accountability for outcomes. Critics of government and public institutions point out that current practice stresses accountability for inputs, which are often very inadequate proxies for outcomes. The criticism raises fundamental questions. What outcomes can be measured? When? Can the outcomes be observed early enough to make improvements? Who is being called to account, anyway?

In the case of NSERC itself, many things can be measured: input variables (e.g., data relating to the NSERC budget and the applications received), process variables (e.g., make-up of Grant Selection Committees, committee workload, referees' reports received, etc.) and outputs in many forms (grant size by discipline, success rates, etc.). NSERC's impact is positive if the applicant and the project are assessed correctly and awarded support to achieve what has been proposed. From that perspective, the published results and the degrees awarded to the research students are the outcome.

But to meet the rising standards of accountability, the measurement has to extend downstream of NSERC, and the research results and the educated people have to be treated as the output. Then the questions are: Have the research results had any impact? Have people noticed them, learned from them, used them, built on them? Have they caused any important outcome in terms of our well-being or prosperity, or in terms of our understanding of nature and ourselves in it? Have they been used in public policy? Have the highly qualified people trained in research gone on to make important contributions? The answers might be available only long after the money is spent, but they will affect future decisions.

Whom will this call to account? NSERC itself, university researchers, industry, or government? Could industry or governments ever be called to account for ignoring research results for which the public has paid? What if they did not know about them? And who is accountable for that? There are practical limits on such questions.

INVESTMENT

It is very important for NSERC that the public should clearly understand what it does. That is why NSERC now identifies itself as the national instrument for making strategic investments in Canada's capabilities in science and technology. The words are carefully chosen; this is not social consumption. But if there is investment, where is the return? The answer is straightforward in project research.[1] The industrial partner puts money on the table because of a prospect of economic return from the project. And it is easy to determine whether a project is successful.

Showing the return on investment in basic research is harder. There is ample historical evidence that many results of basic research have effected profound qualitative changes in our lives. But such outcomes can take a very long time, and they are unpredictable. Nevertheless, for many people they provide sufficient justification for the investments.

But in today's intense competition for scarce public funds, additional arguments are needed, and NSERC is providing them. Together with the research community we are collecting information to demonstrate valuable outcomes that can be traced to basic research supported by NSERC a long time ago. We began this process with successful university spin-off companies exploiting what started out as basic research funded by NSERC. Our first scan identified eighty-two companies, with slightly more than 4,000 employees, and annual sales exceeding $500 million.[2]

INNOVATION

Two kinds of innovation are of particular interest to NSERC: product innovation (the process of bringing new goods and services to market) and process innovation (changing the ways we make or do things). There are other kinds of innovation (marketing and institutional), but research in the natural sciences and engineering produces economic benefit predominantly through product and process innovation.

It is difficult to devise indicators for innovation, a complex process with many feedback loops, involving many institutions and players engaged in various relationships and interactions, and occurring in at least four fundamentally different ways in Canadian industry.[3] However, some rough indicators are used at the level of the firm, for example, the percentage of sales spent on research and development (R&D); the proportion of staff engaged in R&D; the rate of adoption of new manufacturing technologies. All those are inputs, but there are also some output indicators. A good example, which makes sense in the information technology (IT) sector with its very short product cycles, is the fraction of the company's sales that comes from products that did not exist two years ago.

But that indicator does not work in a mature industry where process innovation in making a commodity product is key to staying in business in a price-driven market. In such circumstances, one needs some approach based on value added, or value added per employee, or even "Economic Value Added", net of the cost of capital.

EVALUATION

I think evaluation is not given enough thought or resources in the planning of many programs or policy interventions in the public sector. I have seen studies and reports which are called evaluations, but seem little more than descriptions of how the funds were spent. Better ones have described changes in behaviour. The best have also included indicators showing the progress towards the goals.

Beyond that, there are some fundamental difficulties. First, evaluation is usually premature, significant outcomes of most public programs and policy interventions take too long compared to the mandates of public administrations. Secondly, the evaluation of a policy or program should compare what actually happened with what would have happened without the intervention, just as a result of changes in the environment. However, that is easier said than done. It requires a reliable model to predict what might have happened. I have seen some very elegant theoretical research on evaluating economic policy in this way, but no practical tools.

CONCLUSION

We have to evaluate the performance of the Canadian university research enterprise to improve it continuously and to demonstrate its importance to the public who pay for it. To do this well, an extraordinary degree of interaction and mutual understanding is required among those doing the measurement, those whose work is measured and those who would use the results. It is essential that the appropriate measures be used for all activities, that they be understood in depth and that there be general agreement on what things mean and what they do not mean. This needs to be done well in order to avoid any misinterpretation of the results by policy makers, or any misunderstanding of the indicators by researchers.

Project research has been called "post-academic science"

NOTES

1. Project research was called "post-academic science" by J. Ziman at the conference.
2. "Research Means Business", NSERC, August 1995.
3. "Industrial Innovation in Canada", National Research Council of Canada, April 1994.

CHAPTER 10

The Role of Research Granting Councils from a University Perspective

S. Fortier

INTRODUCTION

In the previous paper, Tom Brzustowski (Brzustowski, 1998) has expressed his view of the role of granting councils from the perspective of someone who lives right at the intersection of science and public policy. I shall take a second look, this time from the point of view of the university whose research life is profoundly shaped by the statistics of federal granting programs.

THE MANY ROLES OF THE GRANTING COUNCILS

Granting councils take on many diverse roles as they interact with different constituencies of the research enterprise, the researchers and research trainees on the one hand, and the consumers and supporters of research on the other. For researchers and research trainees, the councils are at the centre of their research life, playing in turn the role of nurturer, supporter, judge, driver and matchmaker.

ROLES WITH RESPECT TO THE RESEARCHERS AND RESEARCH TRAINEES

Nurturer: Research granting councils all provide significant direct funding to research trainees through programs that support postgraduate and postdoctoral studies. In addition, a large portion of research grants is used to provide stipends to graduate students and postdoctoral fellows.

Indeed, training of highly qualified personnel has become an important factor in the justification of research funding requests and an increasing proportion of research grant budgets. The councils also play an active role in helping Canadian universities attract promising junior researchers through bridging programs as well as retaining them through programs that recognize and reward junior faculty members of exceptional promise. In these capacities, the councils take on the role of nurturer and have a marked influence on career paths and career prospects of junior researchers.

Supporter: At Queen's University, as in most universities in Canada, a large proportion of research funding comes from the national granting councils. This proportion has decreased in recent years, particularly as granting councils' budgets have been reduced. However, this has not resulted in a decrease in the influence of the councils because the remaining part of our research funding pie is not entirely independent of council funding. For instance, several provinces have research support programs with allocation formulae directly based on federal granting council shares. Also, the significant increase in research support from industrial sources experienced in recent years has been largely the result of several new matching programs introduced by the councils to lever funding from industry. Whether one looks at the institutional or individual researcher level, success in granting council competitions is a good proxy for overall success in securing research funding, attesting to the core and vital function played by the councils in supporting the research enterprise in Canada.

Judge: Overall success rates vary markedly among the three federal councils. While the Natural Sciences and Engineering (NSERC) grantees enjoy an overall success rate of almost 80 per cent in the core research grant program, Medical and Social Sciences and Humanities researchers face significantly lower success rates at around 30 per cent. Nevertheless, success in attracting funding from the federal councils has an unequivocal impact on the career progression of researchers regardless of their disciplines. A new grant or a significant increase in funding in a renewed grant will often be one of the key facts invoked when seeking promotion and/or merit pay increase. Holding a federal council grant is regarded as having passed the peer review test. Success in granting council competitions is therefore commonly viewed in the universities as the number one performance indicator of research activities, thus putting the councils in the position of ultimate judge. In this capacity, the councils are seen as having different styles according to their different overall success rates. Because of its much higher success rate, the Natural Sciences and Engineering Research Council may appear to be a more benevolent judge. Its sentences, however, are often terminal and have the potential to put an end to the research careers of academics who fail to attract or retain funding. It is in fact extremely difficult for researchers to restore NSERC funding once they have failed the renewal test, attesting to the fact that NSERC tends to cast its judgement on the whole research career of the applicant. In contrast, it is not unusual to see researchers in the medical or social sciences and humanities disciplines go in and out of funding situations with their respective federal granting councils. While it is more difficult to get an award from these councils, failing to do so carries with it a lower and temporary penalty. These councils tend to cast their judgement primarily on the specific project under review rather than on the whole research career of the applicant.

Driver: Combining a deep respect for the principle of freedom of inquiry with a response to the political pressure to meet national goals and needs has led the federal councils to assume a steering role. A survey of titles found in recent publications of the councils, such as "Partnerships in Knowledge", "Investing in Canada's Health", "Research Means Business", "Investing in Research to Help Build Quality of Life", shows that we are certainly led, if gently, in some clear directions by our research granting councils. More and more researchers get the message that while curiosity-driven research is strongly supported in Canada, it certainly helps if that curiosity is instilled by the desire to answer questions of some national relevance.

Matchmaker: Through many of their newer programs, the councils are enticing researchers to reach out to the outside world and to form partnerships, not only among themselves but also with the private sector. Many new research collaborations have been formed and nurtured through the help of the councils whose role has been similar to that of matchmaker. This has led to a profound change in the attitude of researchers and, indeed, a remarkable change in the culture of research on our campuses. For the most part, these changes have been good, in that they have encouraged academics to broaden their perspective on research and value the full spectrum of research activities from discovery, to development, application and innovation. While in general we applaud the efforts of the councils to help build research bridges, we also recognize the need to support and praise the value of individual research programs, which have an impressive history of important discoveries and innovations.

ROLES WITH RESPECT TO THE SUPPORTERS AND CONSUMERS OF RESEARCH

For the consumers and supporters of research, the councils occupy equally important roles, providing leadership in the elaboration of national research policies, assuming critical accountability functions and occupying a prominent role in the communication of research.

Leadership: The academic research community has long advocated the appointment of a visible and influential champion of research in the federal Cabinet. In the absence of such a champion, the councils and, in particular their presidents, have played a key role. Until very recently, research policies and activities did not appear to secure much time on the agenda of Canadian politicians and policy makers. This situation has changed dramatically in the last year with research issues having a more important presence in both the federal budget and the Speech from the Throne. This can be credited in large part to the leadership of the federal granting councils. The university research community has rallied in support of the councils in their role as promoter and advocate of research in Ottawa, resulting in the elaboration of a clearer and more unified research vision.

Accountability: The financial constraints of the last several years in the public sector have been accompanied by more persistent calls for accountability. Research spending has not been spared from these calls, and the need to assure government as well as the broader public and research community that the research dollars are well spent has intensified. In this regard, the federal granting councils have assumed critical accountability functions, administrating a large proportion of the national research budget. The councils were well positioned to assume responsibilities in this area, having already put in place allocation procedures that are generally viewed as being objective, fair and transparent. They have had, however, to respond to "value for money" accountability requests, a difficult exercise for councils whose goals are intrinsically long term. Harmonizing requests for outcome accountability with long-term research objectives has resulted in intense and, at times, difficult dialogues between the research councils and the academic research community. It has also resulted in a redefinition of research and research activities that is more inclusive and comprehensive. The challenge for the councils is to ensure that all components of the R&D spectrum of activities are, and appear to be, valued and supported.

Communication: Increasingly, the federal granting councils are recognizing the importance of communicating the goals and outcomes of research to the broader public and are putting their efforts into this area. This is apparent in the publications of the councils whose articles are now written to appeal to a broader community. The greater attention to public communication is also apparent in the requests by the councils that researchers make an effort to reach a broader audience, using plain language to describe their research activities. Until recently, one might have described research in Canada as a critical success but a box office failure, so sparse was the coverage of research activities in the public media. This situation has improved significantly in recent years, demonstrating that well-articulated stories can indeed attract the interest of the broader public.

CONCLUDING REMARKS

In a decade of diminishing public funds and demands for accountability, one is being asked for numbers. How many jobs have been created, how many new patents, products and companies have resulted from the national investment in research? These are important outcomes of research, to be supported and reported. But while every one is busy gathering numbers and drawing graphs, is one not at risk of convincing oneself that "that which is most easily counted is that which counts most", as stated by Don Akenson, 1995 Molson prize winner.[1] It is very difficult to find performance indicators that can show the whole picture and accurately report the intellectual, social and cultural wealth created by research. We all have an important role to play in making sure that research, whose outcomes may be beyond what can be easily counted, continues to be valued and supported.

NOTE

1. D. Akenson, Remarks at the Molson prize award ceremony, March 1996, Queen's University, Kingston, Canada.

REFERENCE

Bruzustowski, T.A. (1998), "The View from a Research Granting Council", in *Statistics, Science and Public Policy,* eds. A.M. Herzberg and I. Krupka, Kingston, Ontario: Queen's University, pp. 55-58.

CHAPTER 11

Role of National Academies

C.C. Heyde

INTRODUCTION

National Academies differ greatly in character, charter, scope, size. The perspective I shall give will be on one which is new and small, the Australian Academy of Science (AAS). My perceptions on this subject were forged during nine years on the council of the AAS, including four years on the executive. The AAS is very new, founded in 1954 after Australian Fellows of the Royal Society of London petitioned for a Royal Charter. It has remained quite an exclusive body, currently having some two hundred and ninety fellows. It is growing slowly: initially six new fellows per year were elected; subsequently, this rose to nine, and two years ago to twelve, while about six per year have died.

Unfortunately, the founding fathers and early executive were not very strategic about embracing the applied sciences. Even when nine fellows each year were being elected only one of these was assured for the applied sciences. The result was a breakaway creation, the Australian Academy of Technological Sciences and Engineering (ATSE), which is currently electing thirty-five per year. I believe that it was a strategic mistake to let them go off independently. Governments do not wish to distinguish more than they have to and it has not aided the cause to have two supplicants with two voices.

The government grant to the AAS provides for less than 10 per cent of its operating expenses. Accordingly, it has been necessary for the AAS to be innovative in raising funds. Publications, in particular the high school biology text *The Web of Life* and its successor *The Common*

Threads, have been vital sources of revenue. The Academy also established the Australian Foundation for Science in the late 1980s as a kind of commercial arm. We did not see the major role of the establishing committee as raising the capital base of the Academy, but much more to obtain support from business and industry to fund worthy projects. Many scientific societies have joined the Foundation, as have the majority of Fellows of the Academy. Projects are solicited, funding proposals are written and the Foundation goes shopping for sponsors. In the beginning we hired commercial advice since we lacked in-house expertise. The advice did not meet our expectations and we had to proceed independently. We had difficulty in recruiting an effective chairman from the corporate world who was a significant public figure. Some candidates with experience of fundraising declined with comments like: "Already my friends cross to the other side of the road when they see me coming."

AUSTRALIAN SCIENCE AND TECHNOLOGY INFORMATION SERVICE

Since the Academy was not resource rich, there was a strong desire to use the Foundation to promote public awareness of science. One of the major efforts of the Foundation was to set up an Australian Science and Technology Information Services (ASTIS) in 1988-89. It was a good idea but it was not successfully implemented. It is quite instructive to look at what went wrong.

Now the main features of ASTIS were to provide a media referral service, educational services and services to parliamentarians. It would take me too far off the track to describe the details, but there was much in common with services offered by the Media Referral Service in the United States and the Ciba Foundation Media Resource Service in the U.K. We sought government support for ASTIS under public awareness of science programs and basically foundered over personalities.

Barry Jones was the relevant minister at the time (Minister for Science and Technology). Jones is an Australian icon, an ex-quiz show champion, author of *Sleepers Awake* and currently chairman of the Australian Labour Party. He did not have an easy task in dealing with his ministerial colleagues. However, he was a powerful advocate for science. Indeed, if you asked an average member of the general public in Australia a few years ago to name a scientist there is a significant chance that the answer would have been Barry Jones. On the other hand, he did make the statement that "scientists are wimps" when funds did not come through. This led to an unprecedented public demonstration by scientists at Parliament House.

There were strenuous arguments with Jones, but the funds ultimately went to set up "The Australia Prize", an annual international award. Unfortunately, this attracts little media attention. Some one million dollars per year of public awareness of science-earmarked money is going into this. He did have his successes though. Perhaps the principal one was the 150 per cent tax deduction for R&D spending for corporations. But he was also very helpful to the academy in moves to establish the Prime Minister's Science Council and the position of Chief Scientist. These things have been working quite well and list among their achievements the setting up of a framework of Cooperative Research Centres Scheme worth hundreds of millions of dollars.

AUSTRALIAN ACADEMY OF SCIENCE

But now let me return to the Academy itself and describe some of the problems that it faces. First, the founding fathers of the academy had easy access to government. They could get in to see the Prime Minister at short notice. This facility no longer exists and even access to the Minister is not necessarily prompt. Also, the Minister may not be a member of Cabinet. Government reorganizations are frequent and destabilizing, and it is not easy to get things done. A second problem for the academy is its public profile. A president holds office for four years. They are all very fine scientists, but few have the charismatic touch to become public figures in a real sense. The current President is the exception.

As with all academies, there are concerns about the age structure of the fellowship. One does not want to be seen as a scientific gerontocracy, or indeed to feel like one. The average age at election has crept up from close to fifty to close to fifty-five in twenty years.

The academy has never seen itself as a lobby group as such, and it played a useful role in having the Federation of Australian Scientific and Technological Societies (FASTS) set up in 1985. FASTS has been successful. The academy does work well behind the scenes. There is, for example, an active program of lunches with politicians, key bureaucrats, etc. to foster communication. And the academy is not backward in pressing the political parties at election time, for example, the publication of *Science and Technology: Priorities for the Next*

Australian Government, January 1996, just prior to the last federal election.

What has not been achieved as yet is an umbrella of co-operation between academies such as exists in the United States between the National Academy of Science, the National Academy of Engineering and the Institute of Medicine, that is, the National Research Council. Also, in Australia the Academy lacks any statutory responsibility for the provision of advice to government.

CONCLUSION

All national academies have much to offer their nations, even if their resources are principally human rather than financial. Their real fruits are in what they can foster for science. It is most important that they are not inward looking, despite the fact that they must necessarily spend much time and effort on the election of their members.

CHAPTER 12

The Influence of Statistics in Government Policy Making

E.A. Gehan

INTRODUCTION

I am not in the United States government, but am located in the Capital and have many statistical colleagues at the National Institutes of Health (NIH) or other parts of the government. A number of these individuals have been contacted about current public policies that have been influenced by statistics. From this information, I have chosen several themes to emphasize.

The government is a key element in deciding and implementing public policies. As a biostatistician, I will emphasize the role of statistics in determining public policies. A brief consideration of this topic suggests that "statistics" is not enough. The data must be on your side, but also some advocacy groups that "sell" the policies. Since I work at the Lombardi Cancer Center of Georgetown University, emphasis will be mainly on cancer-related policies. In 1977, M. Schneiderman wrote: "In controversies related to public policy in the prevention and treatment of cancer, the disciplines of statistics and epidemiology have provided the firmest scientific base on which practical political decisions can be made." That is certainly correct, but decisions about public policies are not always decided based upon the best experimental data and the strongest evidence.

INFLUENCES ON CANCER-RELATED HEALTH POLICIES

Policies of government may be built on the statistical evidence flowing from experimental or observational studies. However, there are various other influences which may determine the actual policy. Those in industry, especially the pharmaceutical industry, are concerned with profit margins. The financial support by government for biomedical research in the latter half of the twentieth century has consisted of more support for basic science, clinical trials and epidemiological research, with rather less emphasis on the control of cancer resulting from this research. Recently, however, there has been more support for cancer prevention and the means of control. Patient advocacy groups, especially in AIDS, have influenced policy decisions concerning treatments, in particular acceptance of new AIDS drugs for clinical studies before efficacy has been proven. The cost of a policy is obviously also a major influence. Those engaged in the private practice of medicine have often resented the interference of government in their practices. In the United States in the last twenty years, health maintenance organizations have determined treatment policies by their decisions concerning what strategies of treatment will be supported. Similarly, health insurance companies influence policies by deciding which types of treatment will be supported. At the Georgetown University Medical Center, within the last ten years, the amount of care at the hospital directed by managed care insurers has increased from 5 to 75 per cent. Academic health centres have been placed in the position of competing against hospitals without a teaching or research mission. In a competition based mainly on cost, the academic centres tend to be the losers. Also, the media influences patients about which treatments to seek for their conditions.

Early Example of the Influence of Statistics on Public Policy

In one of the best-known studies of clinical trials, James Lind in 1747 (Lind, 1753) did an essentially randomized study of twelve men aboard the ship *Salisbury* at sea.

> The cases were as similar as I could have them ... they lay together in one place ... and had one diet common to them all. To two of them was given a quart of cider a day, to two an elixir of vitriol, to two vinegar, to two oranges and lemons, and to the remaining two an electuary recommended

> by a hospital surgeon. The most sudden and visible good effects were perceived from the use of the oranges and lemons, one of those who had taken them being at the end of six days fit for duty ... the other was appointed nurse to the rest of the sick.

Though this trial would appear conclusive, Lind continued to recommend other treatments for scurvy, such as "good air". The British Navy did not supply lemon juice to its ships until 1795. The delay in the implementation of the public policy from 1747 to 1795 is typical of subsequent health policies.

The Speeding of New Anti-Cancer Drugs to Market

A headline from a March 1996 article in the *Washington Post* reads "White House Announces Plan to Speed New Drugs to Market: Anti-Cancer Agents Could Get Provisional Approval". How did this new policy get decided? Basically the Food and Drug Administration (FDA) policy concerning acceptance of anti-cancer drugs was catching up with the policy concerning anti-AIDS drugs which had been strongly influenced by demonstrations of AIDS activists. Under the new policy for anti-cancer drugs, effectiveness is demonstrated if tumours are shrunk, whereas previously experimental studies had to demonstrate that the drug was safe and effective. Effectiveness was defined as complete remission of disease or enhanced survival. Ultimately, long-term efficacy studies will still be needed. However, a drug can be provisionally approved based upon its capability for shrinking tumours.

President Clinton announced this policy as an example of the Democrats "reinventing government" and being responsive to the people. However, the Republican's reply was that this policy was recommended some time ago by the "Council on Competitiveness" of the Bush-Quayle administration. Pharmaceutical companies are certainly interested in this new initiative because drugs can now be sold at a much earlier stage. However, a drug that partially shrinks a tumour may not ultimately be effective in providing long-term control of the disease. The head of the FDA, Dr. D. Kessler, reported that no mistakes have been made in eleven accelerated drug approvals so far, but some mistakes may be made in the future. In this circumstance, the data about the effectiveness of the policy are being gathered subsequent to its enunciation.

Evolution of Papanicolaou Test Policies in Cervical Cancer

Breslow, Agran, Breslow, Morganstern and Ellwein (1977) showed that epithelial changes were recognized as a precursor to cancer of the cervix in the early 1900s and Papanicolaou and Babes reported exfoliative cytology as a means of detecting cervical cancer as early as 1928. The discovery of the test by Papanicolaou, a laboratory scientist, was serendipitous. It was a by-product of the study of periodic changes in the reproductive organs associated with the ovulation cycle. Papanicolaou did not seriously pursue his own finding about cancer for more than a decade until he became associated with a gynecologist, Dr. Traut, in the late 1930s. Traut perceived the implications for cancer control and joined with Papanicolaou in perfecting the cytologic test as a practical means for mounting a clinical attack on cervical cancer. Their monograph was published in 1943, but it was not until the 1970s that an estimated 70 to 90 per cent of American women had received this test at least once. In the interim, seven to ten thousand American women were dying each year of cervical cancer.

The current situation concerning cervical cancer screening was given in the *Journal of the National Cancer Institute* (Breen, 1996). The cervical cancer screening rates are substantially short of the national health objective of 85 per cent for the year 2000. Also, the percentages did not improve between 1987 and 1992. The primary reasons given are lack of coverage of the cost of this test by health insurance, lack of a primary care physician and poor communication between women and their physicians. It is unfortunate that the figures on PAP smears are not better, because in April 1996, a consensus conference at the National Cancer Institute concluded that deaths from cervical cancer are preventable.

Smoking and Lung Cancer

Observational and epidemiological studies in the early 1950s by Doll and Hill, Wynder and Graham, and Levin and Goldstein (Schneiderman, 1977) each demonstrated statistically significant differences in the proportions of non-smokers and in the distributions of the amount smoked among smokers between lung cancer cases and controls. In 1957, the Medical Research Council (MRC) in Britain published a statement that "So far no adequate explanation for the large increase in the incidence of lung cancer has been advanced save that cigarette smoking is indeed the principal factor in the causation of the disease." A similar statement appeared in the Surgeon General's report in the U.S. in 1964, that "the principal etiological factor in the increased incidence of lung cancer" is smoking. However, it is only within the last ten years or so that it has not been possible to buy cigarettes at cancer centres or smoke in most workplaces or on airplanes. The tobacco industry still does not accept the major conclusions. This is an example where it has taken a long time between the collection of the statistical evidence and the implementation of public policies concerning smoking.

Studies of the National Center for Health Statistics Influencing Public Policies

Dr. M. Feinleib, former Director of the National Center for Health Statistics (NCHS), recently mentioned several policies that data from the National Center for Health Statistics has influenced. The current policy of having lead-free gasoline was influenced by the National Health and Nutrition Examination Survey (NHANES) of the content of lead in the blood levels of school children. This was one of the forces leading to the Environmental Protection Agency action concerning lead-free gasoline.

A survey of teenage black mothers on welfare established that infant mortality was very high, but that most mothers did not see a doctor before the birth of a child because of lack of funds. This led to liberalizing the eligibility requirements for Medicaid to make it easier for such young mothers to be under a doctor's care prior to birth.

In the 1970s, the speed limit was reduced to fifty-five miles per hour on the nation's highways to conserve gasoline. Surveys of mortality from highway accidents demonstrated that deaths were down 20 per cent over earlier figures. However, very recently this speed limit has been eliminated in favour of having each state set the limits. This was a political decision, mainly by the Republicans, designed to reduce the influence of government on an individual's affairs. The effect on mortality from highway accidents under the new rules remains to be seen.

Collecting vital statistics data certainly makes possible informed decisions about public policies. However, a *Washington Post* headline read "Health Survey's Prognosis is Poor. Data from National Studies Have Guided Policy, Changed the Way Americans Live". In the interest of cutting the government budget, future support for the NHANES will be reduced. C. Everett Koop, former Surgeon General, testified: "It represents a schizophrenic point of view on the part of the government. The greater the truncation of the survey, the less valuable the effort. The bottom line is that this is a bad idea." Currently, the position seems to be that monies might be made available for a reduced cost health survey.

Policies Relating to Prostate Cancer

According to the American Cancer Society, 1995 (Merrill, 1996), the number of new cases of prostate cancer has increased from about 50,000 per year in 1971 to nearly 250,000 per year in 1995, with the major increase having taken place since 1990. The estimated number of prostate cancers during 1995 of 244,000 is higher than both lung cancer (170,000) and breast cancer (183,000). Yet, over the years, the public has heard more about the increase of lung and breast cancer; and research funding for prostate cancer is consistent with the public perception. Though the estimated mortality during 1995 was around 45,000 per year for prostate cancer, breast cancer and AIDS, the annual funding from NIH for prostate cancer is only $59 million, compared with $313 million for breast cancer, and more than $1 billion dollars for AIDS.

Recent trends suggest that federal policies regarding the fight against prostate cancer may change because of various influences. A recent cover of *Time* magazine (March 1996) shows General Schwarzkopf as joining the battle against prostate cancer because he is one of its victims, along with other well-known individuals. M.R. Milken has founded CapCure and is funding a research program of over $20,000,000 per year into cancer of the prostate. With this type of support, it is virtually certain that the United States government will implement increased levels of funding for similar research.

DISCUSSION AND CONCLUSIONS

Recently, there was a review in the *Journal of the National Cancer Institute* of a 1995 book entitled *Cancer Wars — How Politics Shapes What We Know and Don't Know About Cancer* by R.N. Proctor. From this review:

> How often have we identified more knowledge as the key to controlling cancer? From cigarettes to asbestos, from sunlight to radon, from high fat, low fiber foods to radioactive waste, and from pesticides in our food to industrial wastes in our air and water, we have a virtual catalog of cancer causes. But how have we acted on that knowledge? Economic interests and, ergo, political interests, do not want major investments for social policies to control these causes of cancer. It is more politically correct to invest hundreds of millions of dollars to find out how cigarettes cause cancer rather than to invest that money to find ways to decrease cigarette use.

I would like to comment briefly on the role of statisticians in influencing health-related policies. It has been heartening to note that recently statisticians have spoken and written about being active participants in solving problems, rather than either theoreticians or applied statisticians who aid in the design of experiments and analysis of data, but leave it to the subject matter specialists to take the lead on a research project. Mitchell Gail (1996) of the National Cancer Institute chose as the title for his address in 1995 to the American Statistical Association: "Statistics in Action" and used two examples, the first on the randomized controlled trial and the second the debate about whether smoking causes lung cancer and made the point that these examples have certainly increased the awareness of the importance of statistical thinking in medical and public health circles. He concluded by stating that it was "important to seek out new collaborations, to identify good problems, and to steep ourselves in their intricacies. If we do this and collaborate with colleagues in other fields with sympathy, enthusiasm, a deeper understanding and a broader view, then we hope to hear them say 'That statistician really understands our problem and will help us solve it.' That's statistics in action."

The influence of statistics upon health-related policies depends on having experimental and statistical evidence, but there are many other forces involved. Having strong statistical evidence is a necessary, but not sufficient condition for influencing health policies. Of perhaps even more importance is having influential advocacy groups and sufficient finances on your side.

REFERENCES

Breen, N. (1996), "Cancer Screening by Age, 1987 and 1992", *Journal of National Cancer Institute* 88:6.

Breslow, L, Agran, L., Breslow, D.M., Morganstern, M. and Ellwein, L. (1977), "Cancer Control: Implications from its History", *Journal of National Cancer Institute* 59:671-86.

Gail, M.H. (1996), "Statistics in Action", *Journal of American Statistical Association* 91:1-13.

Lind, J. (1753), *A Treatise of the Scurvey*, Edinburgh: Sands, Murray and Cochran, Reprinted by C.P. Stewart and D. Guthrie, eds. (1953), Edinburgh University Press, pp. 145-58.

Merrill, R.M. (1996), "Trends in Prostate Cancer Incidence and Surgery", *Journal of National Cancer Institute* 78:403.

Schneiderman, M.A. (1977), "The Numerate Sciences — Epidemiology and Biometry", *Journal of National Cancer Institute* 59:633-44.

CHAPTER 13

The Waning Role of the State and Science Policy

I. Krupka

INTRODUCTION

Over the last thirty years or so, the role of the state has undergone significant evolution in most Western industrialized countries. This evolution has had a significant effect on the development and substance of public policy, including science policy. A review of some of the features of this evolution can provide a perspective from which to consider the promotion of a productive dialogue between statisticians and scientists on the one hand, and elected and appointed officials on the other. My remarks refer mainly to the Canadian experience, but may have wider validity.

In many Western countries, the public policy response to the depression of the 1930s, the Second World War and particularly the ensuing reconstruction provided much of the foundation for a strongly positive role for the state. Until perhaps as late as the early 1970s, the state was generally believed capable of successfully addressing a broad range of social and economic challenges. This belief was reflected in the wide range of new postwar public policies, programs and institutions, including policies relating to the development of science and technology.

GOVERNMENTAL MACHINERY FOR SCIENCE POLICY

1960s and 1970s

During the 1960s, Canada, the United States, the United Kingdom and Australia, among other countries, undertook major reorganizations of their governmental machinery and developed new processes to formulate and implement policy. The aim of the reorganizations appeared to be primarily to adjust to new and anticipated demands on the state, rather than to prepare for any major change in its role.

In Canada, the development of new, more formal arrangements in the 1960s and early 1970s recognized the substantial increase in the scale and complexity of governmental activity and the related growth and increasing diversity of the bureaucracy. Previous arrangements to develop and implement policy had been relatively informal and relied fairly heavily on personal contact among a small political and bureaucratic elite. Many members of this elite began their careers during the war and were therefore approaching retirement in the 1970s. The theory was not only to reorganize to improve efficiency, but also to provide for institutions capable of serious strategic thought.

The extensive changes to policy making and implementation introduced by Prime Minister Trudeau in his first term (1968-72), created new formal structures and processes which may be viewed as replacing the more informal arrangements operated by the aging elites. The changes did not appear to alter assumptions about the positive role of the state.

Between the mid-1960s and the early 1970s, in addition to many other major changes, the Canadian government established a Science Council of leading scientists which reported to the Prime Minister. A Science Secretariat, headed by a Chief Science Advisor to the Cabinet, was created to report to the Prime Minister through the Secretary to the Cabinet. A Minister of State and Ministry of State for Science and Technology, added in 1971, replaced the Science Secretariat to strengthen the government's ability to develop and co-ordinate policies and programs bearing on science and technology. The Privy Council Committee on Scientific and Industrial Research, a statutory Committee of the Cabinet, chaired by the Minister of Industry, was fairly active during the latter 1960s, as it considered major projects of scientific and technological significance. A major task force was established to review the organization of the government's scientific laboratories, including those of the National Research Council and a Senate Committee reviewed Canada's science policy.

1990s

The completion of the major changes in the machinery and processes of government in the 1970s, coincided, ironically, with a declining popular faith in the ability of governments to meet modern social and economic challenges. Much has been written on this subject, usually spiced with allusions to "the 1970's oil crisis", "globalization" and "the demise of the social contract".

In contrast to the strongly positive role attributed to the state in the period up to the 1970s, governments of all persuasions have been behaving as if the role of the state should increasingly be that of an arbiter among competing economic and social interests, or even simply that of one of the competing interests.

Most governments now make a virtue of reducing the scope of the state's activities. They do this not only in the traditional ways associated with budget reductions, but also in the more fundamental way of questioning the *raison d'être* of specific activities. In many cases, activities are either abandoned or transferred, more or less formally to the private or non-government sectors. Few people within or outside government profess much faith in the government's ability to solve problems.

A significant diffusion of political power from the Cabinet to interest groups, the courts and the media has further reduced the role of the state. In 1989, a senior Canadian minister wondered aloud whether the diffusion of political power had not progressed to the point where the concept of responsible government had become obsolete. How can you hold a government to account for results that it has little or no power to produce?

All of the structures related to the development and implementation of scientific policies and programs which I have mentioned by reference to Canada, were abolished by the early 1990s. Their vestiges exist here and there, largely in the Department of Industry. A Junior Minister, not of the Cabinet, the Secretary of State (Science, Research and Development), who works under the Minister of Industry, was appointed when the government took office in November 1993. An advisory council on science and technology, reporting to the Prime Minister was to be appointed later in the summer of 1996. But these new structures and institutions should not be confused with those bearing similar names created in the 1960s and 1970s.

CONCLUSIONS

The changes I have sketched in the role of the state suggest that the idea of "public policy" as it appears to be used in the name of this conference has a substantially different meaning today from what it might have had as recently as two decades ago. Governments share with other groups the shaping of public policy to a significantly greater degree than in the past, and this is widely accepted as desirable or inevitable both within and outside government. Relying on the creation of new governmental institutions to support science or on finding the right ears at the centre of government to respond to the needs of "science" is not likely to be an adequate strategy to promote the role of science in society. A successful strategy will need to rely much more heavily than in the past on scientists and statisticians proving their utility to society without the benefit of active government support.

PART IV

HAZARDS AND RISKS

CHAPTER 14

The Perils of Poor Risk Management

W. Leiss

INTRODUCTION

This paper presents an overview of analysis and case studies illustrating systematic risk management failures. The consequences of such failures are, among other things: social amplification of risk, difficulties experienced by governments in managing risk controversies, the personal costs experienced by individuals stemming from concern about certain risks, and in some cases significant monetary costs imposed by risk management and risk communication failures. It suggests that at least some of these failures and costs can be avoided if risk experts, who are responsible for characterizing risks in scientific and probabilistic terms, make a more determined effort to understand the expectations which members of the public (who are not generally expert in these matters) have about such things as the timely provision of relevant information, getting responses to the types of concerns that they themselves have (which may be quite different from those of experts), and seeing that experts present technical risk information in a way that is understandable to non-experts.

A DIAGNOSTIC FOR RISK COMMUNICATION FAILURES

The work of risk communication occurs in the gap that separates the expert assessment of risks from the common public understanding of those same risks. Typically, the expert assessment culminates in a technical description of a hazard, using terminology from toxicology or engineering, a detailed account of exposure parameters, and a probabilistic account of the risk to specific exposed populations, with qualifications for uncertainties and confidence levels. On the other hand, the typical "public" comprehension of risk is thoroughly qualitative in nature, involving anecdotal evidence about exposure, a cursory (if that) acquaintance with the hazard characteristics, and emotive expressions of dread, puzzlement, anger and upset, and so forth.

There is no "cure" that will dissolve this gap and render it socially harmless. It will continue to persist and perhaps even to grow, despite the wider circulation of scientific information in quasi-popular publications. It is in the very nature of health risk issues to cause anxiety and concern; moreover, the inevitable uncertainties inherent in risk assessments are exasperating to many people, who think that they have a right to clear and definitive answers.

Where no good effort is made to bridge this gap with credible communications about risks, a risk information vacuum may develop. Such a vacuum may also attract other actors and materials that are not helpful in providing a reasoned social dialogue about risk management. But its worst feature is that public fear, perhaps quite unnecessary levels of fear, can escalate significantly in the absence of a dedicated, long-term effort to explain the meaning of a complex scientific analysis. To give one example, in the case of dioxin risks, it would have been necessary for a competent agency to commit itself to a program of public information lasting possibly thirty years or more.

I postulate that, for governments especially, there is no avoiding the responsibility to make a much better effort in the future to undertake competent public communications about risks of special concern. The main reason is that the costs of failing to do so can be very high.

ANALYSIS

Risk communication failures are a major factor in the difficulties that citizens have in developing sensible attitudes towards risk issues. In brief, most members of the public really cannot figure out what to be worried about and what not, largely because insufficient explanation has been offered to them about the scientific and statistical complexities involved in risk assessments. They also do not have ready access to clear and straightforward explanations of the technical components making up risk assessments, such as:

(i) the ways in which results from experimental studies with laboratory animals are turned into estimated human health effects;

(ii) the very different meaning of the "numbers" used in risk estimates, such as the possible amount of "excess" cancer mortality, on the one hand, and the directly-observable number of "dead bodies" represented in the statistics for such things as traffic accident fatalities, on the other;
(iii) the significance of the uncertainty factors in risk estimates;
(iv) the difference between association and causality in epidemiological studies;
(v) the importance of the difference between hazard and risk, i.e., the exposure factor;
(vi) the conservative assumptions and safety factors built into health risk guidelines; and
(vii) the meaning, in everyday life terms, of very low probabilities.

Many citizens do not understand certain fundamental aspects of risk, again in large part because no decent effort is made, outside formal schooling, to reiterate continually the fundamental attributes of thinking clearly about risks. Some of the commonest misunderstandings are:

(i) thinking that there is a meaningful difference between "artificial chemicals" and "natural substances";
(ii) thinking that zero risk, which also means zero benefit, is better than a risk-benefit trade-off that yields substantial net benefits; or
(iii) thinking that risks are lower than they are actually, just because the technology or activity in question is "familiar" to the individual, for example, automobile driving.

STRUCTURAL DIFFICULTIES

Some mistakes that experts in probabilistic risk assessment commonly make serve to exacerbate the problems citizens have in understanding how to evaluate risks. First and foremost among these is systematic failure to make any decent effort to explain any of the "methodological" matters listed, which apply to all types of risks, in a language appropriate to the knowledge level of the audience. Secondly, and closely related to the first, is the failure to give an account of the scientific research effort lying behind the risk assessments, in those cases where specific substances or effects have given rise to special public fears. It is important to note that no one-time effort would be sufficient to satisfy these requirements; rather, only a sustained communications program, lasting just as long as both the scientific research effort and the issue itself does, in the dioxins case, almost thirty years by now, would adequately respond to the challenge.

Other institutional obstacles intervene as well. First among these is the erosion of trust in major institutions, notably governments and business, especially where matters of environmental protection are concerned. Thus when a risk issue becomes controversial there is a tendency not to believe those organizations which actually have the responsibility to manage risks sensibly on behalf of society. In addition, the structure of news reporting — when things go well, it is not news, but when things go poorly, this becomes newsworthy — highlights the occasional event that may not be representative of the situation as a whole: A minor accident involving a passenger airplane will make the news, but a report that "one thousand airplanes landed safely at Pearson Airport today" will not.

The proclivity of politicians to issue "no risk" or "don't worry" statements ("There is no risk of ... in Canada") lends support to the unfortunate desire of some persons for an "absolute" kind of safety, based on the belief that the risks that worry them can be eliminated, as opposed to being controlled within "acceptable" levels. This is almost always impossible to achieve, except where the benefits associated with certain risk-taking ventures will be foregone entirely. The more common scenario is the substitution of other risks for the one that is eliminated, without those who had the original concerns even being aware that this is occurring. This opens the possibility that risk levels will actually increase, rather than decrease or disappear, as a result of uninformed risk substitution choices.

Perhaps worst of all, so far as the usual public discourse on risk is concerned, is that risk information, including probabilistic expressions, appears sporadically in disconnected chunks, so that meaningful comparisons between technologies or activities are almost impossible to make. Stories appear in the news media about specific events or scientific studies, but rarely are any comparison data given at the time when the reports appear. Moreover, the units of measure vary widely, making comparisons difficult: for example, sometimes "parts per billion" or "a one-in-ten-million chance", sometimes "expected fatalities" (excess mortality, which is difficult to understand). Two essential elements in thinking sensibly about risks are discouraged thereby. The first is *comparative risk* among various components of a related group of risk factors: For dietary intake, the food product label may say "no cholesterol", but what is the fat content? The second is *priority-setting* among a group of intrinsically unrelated risk factors: so far as my children's health is concerned, should I be more concerned about smoking or about trace chemicals in the drinking water supply? The episodic and disconnected reporting of the risk information that reaches most citizens rarely helps them to make these kinds of essential discriminations.

CONCLUSION

All too often, governments, industry and risk experts have assumed that "the public" ought to be able to understand and accept the risk assessments they provide and, when the public appears not to be able or willing to do so (which may have much more to do with distrust), the former conclude that the public is hopelessly incompetent when it comes to rational risk management. This is, however, an unfounded as well as self-defeating judgement. For if it is a matter of lack of understanding, then the onus is on the authors of risk assessments to find a way to make themselves understood. And if it is a matter of the public's unwillingness to accept expert judgement, no matter how much effort has been put into overcoming insufficiencies in understanding, then it is a matter of the fundamental right of citizens in a democratic society to have the final word in the matter and, of course, to take the consequences if those ultimate choices should prove to be unwise or disadvantageous to themselves.

Very often it will be found, in cases of public risk controversy, that there are "extraneous" factors, intrinsically unrelated to the specific risk issue itself, that have come to play a major part in the failure to reach a reasoned consensus. Some of these commonly-seen factors are:

(i) key information is not revealed, or is revealed only under duress, or not in a timely fashion;
(ii) information is incomplete or misleading;
(iii) information appears to be internally contradictory;
(iv) equally-qualified experts appear to be taking different positions;
(v) those with economic interests in the outcome appear to be in control of the information flow; and
(vi) the public is told that they just do not understand the technical information.

Secondly, in many cases the dispute appears to be about the nature of the technical risk information itself, whereas it is really about differing *meanings* that the information can have with respect to the personal situation of individuals who think that they will be affected by what is proposed. People fear some things more than others, a fear that is not captured in the dry statistics of comparative fatality risks. People have much higher concerns about things that appear to imply increased risk for children, and, if the population-average risk statistics are not disaggregated to deal with subcomponents of the risk profile, they may resent the failure to do so. Perceived unfairness in the distribution of risk across the population in general is also always something that will not escape attention, which makes facility siting controversies, for example, especially bitter. And there are many other "subjective" aspects to the provision of risk information which, if ignored, will surely come back to haunt the unwary technical risk assessor.

What can we do to improve matters? Obviously this is the type of question that cannot be answered adequately here, although a few indicators can be given as to the directions in which answers may be found. These indicators may be outlined in the form of prescriptions to risk experts:

- (i) make the communication of probabilistic estimates as accessible as possible, in terms of the ways in which the non-expert public thinks in everyday life situations;
- (ii) make an effort to understand the terms in which the intended audience is framing the risk issues, so that your information addresses their concerns, rather than yours exclusively;
- (iii) tell the whole truth and nothing but the truth when communicating about risks, including uncertainties, and do so voluntarily and in a timely fashion;
- (iv) avoid "there is no risk", or "there is nothing to worry about", statements.

These prescriptions are accompanied by two things to avoid: first, do not blame the media for allegedly "misrepresenting" the facts; people trust the media more than they do experts who work for government or industry. Secondly, do not blame the public or citizens themselves for being allegedly "irrational" about risk-management judgements, for this will only make the ultimate objective of sensible decision making that much harder to achieve.

CHAPTER 15

Problems in Correctly Interpreting Hazards and Risks

D.V. Bates

INTRODUCTION

The problem of the correct interpretation of epidemiological data is not often one that involves disagreements about mathematical issues. The interface between the investigators and the media is made complex by three main factors, which I will discuss in turn.

HAZARD IDENTIFICATION

The question here is whether a hazard has been correctly identified. The most difficult cases are those in which, from animal work, there is a strong presumption of hazard, but there is no human epidemiology to indicate that it actually exists. This is why dioxin is a difficult and still contentious issue. This situation is not common, however, and more often the existence of a potential hazard is admitted, as in the case of lead, but the question is rather at what level of exposure there is a significant risk of adverse effect. This discussion often begins with the question of whether a causal inference can be made from a statistical association.

CAUSAL INFERENCES

Sir Austin Bradford Hill (1965) was not the first person to consider the question of causal inference; but he was the first epidemiologist or biostatistician to give the question careful consideration. He suggested several criteria which he found useful in reaching a causal inference, and these have been the subject of much recent elaboration. We can recognize (and it is important to do so) that we are all biased in one direction or another when required to come down on the side of a causal inference or to deny it. I have listed some of these biases within my own experience (Bates, 1994); there are interesting reasons why civil servants in the parliamentary system are biased against making a causal inference, but in the U.S. system of government they may be biased in favour (it may be possible to extract more money from Congress). The public in general is biased in favour, particularly if there is a possibility of harm (as, for example, in the case of childhood leukemia and proximity to electric transmission lines); and this is also true of juries. I have noticed that animal toxicologists, trained to devise experiments with the maximum limitation of variables, are very reluctant to make a causal inference from an association.

The media have difficulty with this problem, but it is not theirs alone; and I have to record that Sir Ronald Fisher did not believe that the association between cigarette smoking and lung cancer was causal. What is needed is replication of epidemiological work to test the consistency of the association that has been found, in a different population or geographic region. I have also argued that "coherence" of adverse health effects is an important property of data relating such outcomes to air pollution.

These problems are recurrent and intractable, and there does not seem to be any easy way around them. One criterion is "biological plausibility", and some scientists are unwilling to make any causal inference unless the mechanism for such an effect is understood. This seems to me to be too extreme a position, as it did to Bradford Hill.

In a recent British document from the Ministry of Health on air pollution from particles, the conclusion that the association between such pollution and daily mortality was causal was expressed as a double negative: "It would be imprudent not to conclude...."

MULTI-FACTORIAL DISEASE

Most modern diseases, such as heart disease, asthma, lung cancer and others are "multi-factorial". That is, their development is a consequence of the addition of a series of risk factors, some of which may be genetic in origin, and others of which may be environmental. It is very difficult to think straight about "causes" in such instances, because the factors do not add up to one hundred if they are multiplicative. Thus, if workers exposed to asbestos did not smoke, you would prevent 92 per cent of the excess rate of lung cancer; if they smoked but were not exposed to asbestos, you would prevent 81 per cent of the excess rate (not 8 per cent as you might have thought). You sometimes hear someone say: "If 90 per cent of the cases of lung cancer are due to smoking, that only leaves 10 per cent for everything else, does it not?" This is, of course, quite false. It is not surprising that the media and the public have difficulty with this, as everyone does until they are forced to consider the complexity of multi-factorial disease.

This problem also occurs, as it were, in reverse. Recently the mortality from heart disease has been declining in many countries. Is this due to better acute medical care? Or to changes in diet? Or to jogging? Or to open heart surgery? Possibly all of these have played a part, but we would still like to be able to apportion the credit between them.

EPIDEMIOLOGY BASED ON LARGE BANKS OF DATA

Modern computer technology has enabled very large banks of data to be analyzed. In the field of air pollution epidemiology, very large sets of data are now available. One study from Ontario analyzed hospital admissions from one hundred and sixty-eight hospitals over a period of six years (Burnett, Dales, Raizeen, Krewski, Summers, Roberts, Raad-Young, Dann and Brooke, 1994); it found a strong association between admissions for respiratory disease and air pollution levels. Another study involved a seventeen-year follow-up of over 555,000 Americans, including their smoking habits, etc., and indicated that survival was related to the air pollution level of the city in which they were living (Pope, Thun, Namboodiri, Dockery, Evans, Speizer and Heath, 1995). This kind of data immediately raises the question of whether the association might be confounded by undetected factors that differentiate the populations, and it is not surprising that the media have difficulty with that kind of question. Nevertheless, the very large numbers involved in such studies give them statistical power which was not possible before.

PHRASES TO NOTE

In the public debate of many of these issues, certain phrases are used so commonly, and are so often out of place, that it is worth listing them. They include: "No scientific proof that ... "; "No statistical proof that ..."; "No significant excess of ..."; "No objective evidence of ...".

Whenever one of these phrases is used, the speaker must be interrupted to explain what, in his view, would constitute such proof in relation to the question at issue. Proof in the mathematical sense is rarely possible in biology.

QUANTITATIVE RISK ASSESSMENT

When quantitative risk assessment was first introduced, there was hope that it would short-circuit much of the "political" bias introduced into arguments about environmental health hazards. As Silbergeld (1993) has pointed out, it has largely failed in this objective. The reason for this is that there is bound to be a great deal of judgement exercised in relation to the input data; and consequently the upper and lower bounds of such risk estimates are often too wide to be of much use. Nevertheless, it does have a useful part to play in decision making, and in some instances, such as balancing cancer risks, it has proved to be very informative. One of the major difficulties is to extrapolate animal data to the human. And this is bound to involve many uncertainties. This difficulty is often obscured when detailed human cancer risks are presented.

DECISION MAKING IN FREE SOCIETIES

Observing the history of decision making in the West over the past forty years, it seems to be true that, as evidence accumulates, there is a subtle shift in the burden of proof. Thus, as the evidence against cigarettes began to accumulate, the burden of proof shifted to the tobacco industry to explain why the epidemiological data were so consistent in so many countries; we have now reached a situation in which only the beleaguered executives of the tobacco industry are prepared to state that: "There is no scientific proof of a causative link."

How do we decide that the time has come for legislative action? Here the role of the media is crucial in forming a public consensus that the hazard must be controlled. It is interesting that the public involvement on such decisions seems to be variable; thus it has been strong in relation to air pollution and asbestos; but almost absent in the decision to take lead out of gasoline. The attention of the media is therefore an inconstant factor, but it remains a very important one. It follows that all students in environmental sciences should have some introduction to speaking to the media on controversial issues. It is, for example, important to recognize that when a member of the media asks whether air pollution is a more important issue than something else, he has to be reminded that there is no generic "order" of importance. If you live in a region of high particulate pollution, it may be specially important; but elsewhere possibly the prevention of flooding or the maintenance of firefighting surveillance might be a higher priority.

CONCLUSION

Sometimes it seems that our agenda is being driven by the "environmental fringe" and the continued obsession of the media with their expressions of concern. The public has a hard task in steering a sensible middle-course between the anxieties frequently expressed in the media, and the discredited denials of government, or the vested interest statements of industry. There is no doubt that the proper management of environmental concerns and the avoidance of preventable pollution both require an informed public, if remedial efforts or inconvenient regulations are to be accepted. This is particularly the case in relation to air pollution, where an informed public opinion has, in most cases, pushed the reluctant politicians towards higher standards and better controls. The fact that gross air and water pollution was much better managed in the industrial West than in the communist world, is striking confirmation of the need of an open society, unrestricted data access, and an informed public opinion in the management of these problems.

In a "free" society, the media play a crucial role in drawing attention to hazards, in publicizing the scientific and political debate in relation to them, and in explaining the impact of proposed regulation. Scientists and epidemiologists have an unavoidable responsibility to work with the media, even if they are sometimes misquoted or their views are misrepresented.

REFERENCES

Bates, D.V. (1994), *Environmental Health Risks and Public Policy: Decision-Making in Free Societies*, Seattle and London: University of Washington Press.

Burnett, R.T., Dales, R.E., Raizenne, M.E., Krewski, D.W., Summers, P.W., Roberts, G.R., Raad-Young, M., Dann, T. and Brooke , J. (1994), "Effects of Low Ambient Levels of Ozone and Sulphates on the Frequency of Respiratory Admissions to Ontario Hospitals", *Environmental Research* 65:172-94.

Hill, A.B. (1965), "The Environment and Disease: Association or Causation?", *Proceedings of Royal Society of Medicine* 58:295-300.

Pope, C.A. III., Thun, M.J., Namboodiri, M.M., Dockery, D.W., Evans, J.S., Speizer, F.E. and Heath, C.W., Jr. (1995), "Particulate Air Pollution as a Predictor of Mortality in a Prospective Study of U.S. Adults", *American Journal of Respiration and Critical Care Medicine* 151:669-74.

Silbergeld, E.K. (1993), "Risk Assessment: The Perspective and Experience of U.S. Environmentalist", *Environmental Health Perspectives* 101:100-104.

CHAPTER 16

Scientific Evidence and Uncertainty in Public Inquiry

J.F. Lawless

INTRODUCTION

Science and statistics play important roles in those branches of public affairs in which decisions or recommendations must be made.

In many situations there is considerable uncertainty about the scientific evidence or its implications. This raises issues about the roles of science and scientists, and the ways in which uncertainty in scientific evidence is handled and communicated. If there is a message, it is that the many contributions made to public inquiry notwithstanding, scientific communities should be more engaged in public affairs and should better educate both scientists and the public about the role of science in these matters.

SCIENTIFIC EVIDENCE IN PUBLIC INQUIRY

Scientific evidence is presented in various ways. In court cases or formal inquiries, an individual expert often presents evidence in association with a particular position or side. Testimony sometimes focuses on rather precisely defined questions, such as: Are the state lottery's published odds of winning the Jackpot Prize correct? In other inquiries, the expert may be associated with a particular side, but the question is less well-defined. Such as, does the fuel additive MMT affect emission-control systems of cars? Do silicone breast implants cause atypical connective tissue disease? Finally, there are situations where panels of scientists aim to provide an objective, or at least a broad-based assessment of evidence, for example, on the effects of second-hand smoke on the health of humans.

There are several aspects of public inquiry that are important to recognize. First, science and reason are but two components of the process of public inquiry. Indeed, plain ignorance along with the self-interest of individuals or groups may often subvert science. In a recent speech to the American Association for the Advancement of Science, U.S. Vice-President Al Gore noted how, two weeks after the Nobel Prize in chemistry was awarded to scientists for their work in ozone depletion, Texas Congressman Tom Delay stated: "The science underlying the CFC ban is debatable" and, he maintained, the agreement to terminate the use of CFC's was "the result of a media scare." There have been several discussions in the media recently about "junk" science. In an article in the *Globe and Mail* (December 21, 1996, p. B2), Terence Corcoran discussed encouraging developments in the U.S. judicial system, wherein standards that judges should insist on when dealing with scientific issues have been specified. These include the principles that "... (scientific) theories must be generally accepted by the scientific community, subjected to peer review, supported by scientific reasoning, backed by reliable data, and reached through scientific method." He went on to note that in a current case before the courts, evidence about the health effects of silicone breast implants failed dismally to meet these principles.

Secondly, it is also important to recognize that, even when there is reasonable agreement about the scientific facts, perceptions of risks or of the benefits attributed to specific actions or events may vary widely across individuals or groups. As a result, a consensus about decisions or recommendations may be difficult to obtain: areas such as health policy and environmental assessment run into these dilemmas head on. Finally, few decisions are made in isolation. It is necessary to understand the systems of which the immediate concern is a part and to consider the effects of actions on those systems.

Scientists must communicate with the public in plain language. Complicating the picture is the fact that scientific evidence often has considerable uncertainty associated with it. I shall turn now to how this uncertainty is expressed and handled.

DEALING WITH UNCERTAINTY

Uncertainty has several forms and sources. One is a simple lack of scientific knowledge about a process or phenomenon. For example, we do not know which aspects of climate change are predictable. A second type of uncertainty is due to inherent randomness in a process: for example, we do not know how many new persons will acquire the Human Immunodeficiency Virus (HIV) through sexual activity in the next year, or how many hurricanes will threaten the coast of Florida. Scientists should distinguish, however, between processes that are sufficiently stable and well understood that statistical predictions are possible, and those that are not. Finally, there may be uncertainty about a phenomenon because we are able to observe only a portion of the relevant universe. For example, we rely on polls of 1,000 or so individuals to provide estimates of public opinion in large populations.

Scientists can thus ask themselves: What is known? What is uncertain? What is stable and statistically predictable? What is not predictable with any degree of accuracy? For the public and for policy makers, uncertainty may also arise from ignorance, from an inability of scientists to convey information clearly, or through conflicting evidence from scientific experts. In adversarial proceedings where experts often seek to downplay uncertainty, the public may need to assess the credibility of the experts as well as that of the scientific evidence.

In situations where uncertainty is primarily due to well-understood statistical fluctuations, there are straightforward probabilistic methods of handling and expressing it. As a rule, it is wise to bear in mind that biases in data and inappropriate scientific assumptions tend to dominate sampling error in most situations.

Scientists should recognize uncertainty explicitly, because the public is easily misled into thinking that there is little uncertainty, and may often be manipulated as a result. Furthermore, great pains should be taken to convey clearly scientific evidence, including uncertainty, to policy makers and the public.

Many scientific panels take a realistic view of uncertainty, even when policy makers might prefer a more precise assessment. Reports of working groups in the U.S. and the U.K. in 1988-89 which forecasted the number of future AIDS cases make interesting reading. They recognized the limitations about the existing information on HIV infections and the progression from infection to AIDS, and consequently made forecasts under a variety of models and assumptions. This produced a very wide range of estimates for future AIDS cases; for example, forecasting from 1988, the U.K. group obtained estimates for the number of AIDS cases by the end of 1992 that ranged from about 9,000 to 34,000. Such wide estimates may be viewed as scientifically weak by some non-scientists. However, an advantage of panels is that individual scientists tend to downplay uncertainty; a display of variation in individual panel members' assessments can better indicate the actual uncertainty. In fact, the numbers of AIDS cases occurring by the end of 1992 turned out to be just under the lower limit of 9,000.

It is also important when revisiting past events, as many inquiries do, to recognize the uncertainties that existed at the time. The Krever Inquiry into actions of the Red Cross and others associated with the Canadian blood supply in the early to mid-1980s has no doubt had difficulty assessing the effect of uncertainties faced by decision makers during that period.

If policy makers and decision makers are to have a clear picture of the uncertainties inherent in a situation, it is the duty of scientists to convey these uncertainties. In settings where scientific knowledge is limited, the importance of system design and adaptability should also be emphasized, particularly when a process or system is unpredictable or where certain events may have very severe consequences. For example, in order to certify a nuclear facility as "safe", it is important to confirm that the facility has the ability to react quickly to critical events such as pump shutdowns in cooling systems. I was interested to hear recently that the University of Western Ontario's School of Business has decided that its students should be told that not everything is analytically or scientifically predictable, and has decided to teach what they term "intuitive" skills for reacting or adapting rapidly to change or unforeseen events.

I shall not say much about the mechanics of communicating uncertainty. Both the way that uncertainty is quantified and the way that it is communicated non-technically need careful consideration. In some areas such as the reporting of polls in the media, conventional procedures have been adopted which many people come to understand over time. Similarly, in risk assessment a great deal of thought has gone into how best to communicate risks. The Royal Society of London's 1992 report, *Risk: Analysis, Perception and Management,* makes clear the importance of recognizing emotional, psychological, and other factors associated with the target audience.

EDUCATION

Science programs at most universities do not prepare students to deal with the issues I have been discussing. Even statistics ignores them almost entirely. Students are usually taught large doses of statistical technique, computing and mathematics, but little about sources and consequences of variation, the different aspects of uncertainty, public affairs, or what one might term "systems thinking", including the ability to think about the ways that specific factors or actions fit into a system. Perhaps our students are buried under an increasing pile of fact and technique which makes it ever more difficult for fundamental concepts to be assimilated.

Any scientific education should arguably include: (i) the logic and practice of both deductive and inductive reasoning; (ii) fundamental statistical concepts such as variation, correlation, conditioning and the planning of studies; (iii) methods of expressing and handling uncertainty; (iv) systems thinking; (v) the role of science in public affairs; and (vi) the communication of scientific evidence, including uncertainty.

There is also a need for a greater synthesis of science, statistics and socio-political ideas. This has been discussed a good deal in areas such as risk management, where the difficulties of balancing scientific knowledge, uncertainty costs and attitudes to risk have been studied. The Royal Society report mentioned that:

> The notion that risk assessment and risk management are overlapping, but separate, tasks is one that scientists and technologists often find attractive, but ... it is too simplistic and misleading.... Indeed the nature of the appropriate structures, their purpose and, especially, their design are the subject of a debate.... These debates have as their significant underlying issue the need to bring together natural science expertise and knowledge about human behaviour and the operation of human institutions (Royal Society, 1992, p. 7).

This suggests that scientists should not merely present scientific evidence, but should also play a broader role in policy making and decision making.

I would add that in addition to teaching scientists about how to be effective in public affairs, we must educate the public about science. A better public understanding of science can only help to make policy making and decision making more informed.

CONCLUSION

Science and reason must compete in public affairs with economic considerations, vested interests, political agendas, and public attitudes and perceptions. We owe it to society to see that scientific evidence is presented clearly and objectively. We will be more effective if scientists are better educated about public affairs and if the public is better educated about science.

Many scientific societies are increasingly emphasizing public affairs. In my own area of statistics, the American Statistical Association has an Office of Scientific and Public Affairs and sponsors a science-writers program. Sadly, such activities are not without opposition from some members of the discipline. Looking outward does not compromise the purer research activities of a discipline, and it offers increased opportunities for communication and for science to argue for public support.

I believe there is still much scope for improvement on the part of educational institutions and scientific organizations. We should work to foster an interest in public affairs; to educate both scientists and the public about topics such as the role of science and the effects of variation and uncertainty in public policy and decision making; and to promote system-thinking and multi-disciplinary study of important issues.

Finally, scientists must impress upon government the need to structure the processes of public inquiry so that sound scientific evidence may be presented, and promote the formation of panels that provide a balanced assessment of scientific evidence. Given the numerous competing constituencies in public affairs, we should never cease to make the case for science and reason.

REFERENCE

Royal Society (1992), *Risk: Analysis, Perception and Management*, Report of a Royal Society Study Group. London: The Royal Society.

PART V

STATISTICS

CHAPTER 17

The Importance of Statistics in the Gathering and Analysis of Data

D.J. Hand

INTRODUCTION

In order to develop public policy, a government must have available facts describing its country. These facts provide the input for feedback mechanisms, so that one can judge if the policy is successful and act accordingly. For this process to have a chance of being effective, the facts must be valid and accurate. Since countries are societies of people, the facts will inevitably be statistically based. The situation is complicated because what is being measured will usually be some kind of construct which may well not really measure what one is actually interested in. The latter may defy clear definition. Crime statistics provide an illustration. We are probably really interested in questions such as: Is the country becoming safer or more dangerous? Are the police doing a more effective job, or are they failing to keep up with things? What are the most effective strategies for reducing crime? To answer such questions we count numbers of reported crimes, numbers of arrests, numbers of arrests which lead to a conviction, and so on. But these are simple measures arising from a complicated situation. Perhaps whether or not a crime gets reported depends on the perception that it will be solved. An increase in the number of arrests could mean that the police are increasing their efficiency or it could mean that the crime rate is going up. A further complication is that groups of people, unlike nature, are not blind systems. Once it has been decided what measuring instrument to employ, then optimizing that measure becomes the objective. If number of arrests is the measure, then apparent improvement arises if one arrests more people, regardless of whether this means more criminals are being caught.

ACCURACY OF CONCLUSIONS

It is also important to be aware of limits to the accuracy of the conclusions. Election opinion pollsters are very careful to place error bounds on their conclusions, but similar issues apply everywhere. We, and also the general public, must be able to assess the confidence which should be put in the conclusions.

Statistical methods are used to answer questions. But, to obtain correct answers, we must be clear about what the questions are. In particular, questions relating to individuals and superficially similar questions relating to society may differ. To illustrate, both medical diagnosis and medical screening require the classification of people into disease classes, a fundamentally statistical operation. However, the first relates to individuals (Does this particular person have the disease?) while the second relates to groups of people as a whole (Which part of the population is likely to have the disease?). Which one is appropriate will depend on the objective of the study. Other such illustrations abound, and the essence of all of them is that there may be a conflict between what is appropriate for the individual and what is appropriate for society. This is a fundamental conflict, and not simply an artefact. It means that, in attempting to discover "what is best" one must be very clear about for whom or for what it is intended to be "best".

Questions about society are questions about groups of people. As such, they inevitably require statistical answers. Questions about individuals are answered by calculating probabilities for those individuals, again based on summarizing groups of people using statistical methods. The difference lies in whether one applies the question to the summary or summarizes responses to the question.

THEORIES AND SCIENCE

At a higher level still, our theories must match the data if they are to have any practical value. But this is not always remembered. The Nobel Prize winner Wassily Leontif analyzed articles from the *American Economic Review* from March 1972 to December 1981 and found that over half were "mathematical models without any data". He went on to say that "year after year economic theorists continue to produce scores of mathematical models ... without being able to advance, in any perceptible way, a systematic understanding of the structure and the operations of a real economic system."

And at the opposite, very lowest level, fraud can occur, not just in the data, but at the design stage, and at intermediary and end-points of a statistical study. Indeed, scientific fraud, though apparently blessedly rare, has become the focus of much attention recently, with some high-profile incidents. Double blind studies in medicine are not effective merely for eliminating subconscious bias. Of course, for science this is probably not a serious issue: science is a self-correcting process. A mistaken idea, whether due to dishonesty or to a mistake, may survive for a while, but it is the very essence of science that it examines itself, seeking and correcting weak points in its structure. The same is not true of public policy, which is built on foundations that necessarily have to be assumed to be solid. If the foundations should change then the rationale for the public expenditure, the justification for doing things the way they are being done, vanishes. The situation is probably even worse if the scientific foundations are not yet established — if they are changing even as we speak. The case of Bovine Spongiform Encephalopathy (BSE) in Britain springs to mind. Science has suffered a lot of criticism recently because of what is seen as its inability to make definitive statements about any link between BSE and Creutzfekdt-Jakob Disease (CJD). But these criticisms reflect a lack of understanding of the scientific process, of probability, and of statistical methods and arguments. One cannot help but note that, if the majority of the population lack this understanding, including, apparently, the leader writers of our major newspapers, then scientists and especially statisticians are in a vulnerable position.

CONCLUSION

I suppose this is a plea for more education in basic statistical concepts. Not so much in numeracy, others are pleading for that, but in such concepts as variation, probability and representativeness. At the very least this would help to quell what some people see as an increasing tide of the irrational sweeping over society, manifest in such things as apparent beliefs in the claims of astrology and schemes for picking winning combinations of numbers in the national lottery.

CHAPTER 18

Economic Arguments for Public Funding of Research: The Case of Economic Modelling

P.B. Dixon and D. McDonald

INTRODUCTION

Our area of research is economic modelling, particularly computable general equilibrium (CGE) modelling. The main strength of CGE models is in giving insights on the likely effects of disturbances in one part of the economy on activity in other parts.

In Australia, CGE modelling has been particularly useful in the tariff debate. After the Second World War, Australia developed its manufacturing industries behind a high tariff wall. This left Australia with expensive, inefficiently produced manufactured products and a generally lower standard of living than otherwise. The problem was how to reduce protection. Many feared that the result would be sharp increases in unemployment.

Via CGE modelling, it was possible to provide evidence that reductions in protection would have little impact on aggregate employment. The linkages in our CGE model traced the effects of: (i) reductions in protection through to reductions in prices in the Australian market of both imported goods and Australian-produced competing goods; (ii) reductions in these goods prices through to reductions in wage demands; and (iii) reductions in wage demands through to increased employment opportunities in export-oriented industries which depend on cost competitiveness in international markets.

The result indicated by our model was that employment losses in import-competing industries were likely to be offset by gains in export-oriented industries. This was helpful in reassuring the Australian public that protection could be reduced without adverse short-run consequences. Protection has now been largely eliminated.

Like all researchers we think that our field is important and deserves public funding. The remainder of this paper outlines general arguments in support of public funding of research and discusses their application at the Centre of Policy Studies (CoPS). These arguments are derived from economics. An effect of economic rationalism is that scientists need to produce convincing economic arguments to justify public financing.

WHY SHOULD THE PUBLIC FINANCE RESEARCH?

Arguments to support public funding of research can be grouped under four headings.

Public Goods

The defining characteristic of a public good is that members of a community cannot be excluded from enjoying benefits from it. An example is space exploration. We can all enjoy the achievements of space programs without making any financial contribution. Because people cannot be excluded from the benefits of public goods, these must be government financed.

To make a public-good argument for government support of their research, scientists must show that (i) the products of their research cannot be sold because of lack of excludability, and (ii) that the community places sufficient value on these products to be willing to pay for them *via* taxation.

Apart from space exploration, other areas in which public-good arguments can be made include greenhouse research, research on the origins of the universe, archaeology and analysis of the arms race.

High Up-Front Costs

We have not included in our list of research-based public goods the outputs of medical, agricultural and geological research. Their products are potentially providable by market mechanisms. Nevertheless, scientists in these areas may have a claim to public funding because the market mechanism, although available, may be highly inefficient. Market mechanisms are inefficient when there is a large up-front development cost and low marginal costs.

In areas of science yielding building blocks for other scientific endeavours, the market-inefficiency argument is persuasive. In these areas, results are obtained at high cost. Although it might be possible to pay for the basic research through licensing and other commercial arrangements, such arrangements reduce the benefits to society. To achieve the full benefits of fundamental discoveries, society must provide public financing of the underlying research, allowing free dissemination of the research results.

Externalities

The externalities argument is similar to the public-good argument. In the case of externalities, part of the product is a private good, that is, one that can be sold *via* the market and part of it is a public good, the externality.

In science, an externality argument for public funding is that scientific research provides training. Inexperienced researchers may, because of labour laws, be paid more than the value of their contributions to research programs. However, the benefit to society of their employment may exceed the required wage payment. This is because young research workers receive training which can be used in future research projects, possibly with other employers. In the absence of public subsidization of the training component of employment, there is likely to be under-provision of research training.

Difficulties in Protecting Copyright

An argument for public funding of research can be made if there is a need for public exposure of the details of discoveries and these discoveries can be copied easily. In the social sciences, there is often a need to expose data collection and estimation methods to detailed peer review. Once exposed, these can be copied with sufficient variations to make copyright protection impossible. Thus, left to market forces, research of this type is likely to be under-performed.

APPLYING THE ARGUMENTS FOR PUBLIC FUNDING OF RESEARCH AT THE CENTRE OF POLICY STUDIES, MONASH UNIVERSITY, AUSTRALIA

In 1995 the Centre of Policy Studies employed twenty people. The sources of its funds were: sales of private goods, 35 per cent, for example, forecasting services; provision of University services, 20 per cent, for example, undergraduate teaching; and government-funded research and training activities, 45 per cent. Over our twenty-one-year history, the arguments that we have found to be useful in persuading the government to continue its funding have been increasingly economic.

We have argued that our work produces public goods in the form of analyses which improve the conduct of economic policy in Australia. These improvements can be enjoyed simply by living in Australia.

We have argued that our research products have high up-front costs and low marginal costs. Our research output is economic models that are costly to construct. Once constructed, these models can be applied by others at low cost. Thus, to ensure an optimal level of use, the models should be supplied to the public at low cost.

We have emphasized our role in producing externalities *via* the training of inexperienced research workers.

Perhaps the most important argument justifying public finance of research in our area is the difficulty in protecting copyright. Economic models are largely about conditional forecasting. They answer questions of the form: if the government adopts policy x, and other events a and b happen, then the outcome for Gross Domestic Product will be z. Events a and b usually do not happen and there is no way of conducting controlled experiments in which they do. Consequently it is hard to assess economic models by their forecasts. Thus, the main method of assessment must be peer review. Because economic models are used in politically sensitive decision making, reviews of their quality should be open to public scrutiny; and because they are easily copied with small variations, open review processes are incompatible with copyright protection.

CONCLUSION

In Australia, applications for public funding of research need more rigorous economic justifications than they did in the past. Because society now doubts the ability of governments to pick winners, there is a preference to leave the bulk of research funding to the market place. Researchers seeking public funding must not only demonstrate the likely benefits of their projects, but must also argue that the public purse rather than the market is the appropriate funding source. This requires arguments of the type outlined in the second section and applied in the third.

PART VI

SCIENTISTS AND STATISTICIANS

CHAPTER 19

Statistics in Science

Sir David Cox

INTRODUCTION

I had some problems about what the title actually means. I have taken it in the following spirit: the clear successes of statistics are in the science-based technologies, in medicine (in the randomized clinical trial and in the careful epidemiological study), in agriculture (in field trials and programs of plant breeding, and so forth), and in industrial quality control, reliability and associated areas that are developing so excitingly. The role of statistics in science, contrasted with science-based technology, is perhaps less clear.

THEORY OF PROBABILITY AND STATISTICAL ANALYSIS

I prepared this paper on the basis that most of the readers will not be professional statisticians. Therefore, I thought it might be useful to give a summary of the subject as it seems now. We may start from two fields: pure mathematics and the philosophy of science. The theory of probability, of course, began with games of chance, which have just possibly no interest to serious-minded people. But it has developed into a most beautiful and active area of pure mathematics with links to number theory, potential theory, functional analysis, and so forth. Theoretical probability leads on to applied probability, which is basically the applied mathematics of situations in which probability is the essential element. Such problems arise, for example, in physics, in a number of areas of biology, in economics, especially finance, and to a limited extent in sociology.

The theory of probability is also needed for the theory of statistical inference.

Philosophy of science, as discussed by physicists, is serious. What the philosophers have to say, perhaps, is something rather different, but the conceptual role of probability in studying uncertain inference is important here.

The principles of design and analysis of investigations form the core of statistics. They are formulated mathematically, but many are basically qualitative concepts.

Let me give just two examples of applied probability and statistical analysis in the physical sciences. Think of a long river. It will have tributaries. Those tributaries will have sub-tributaries. In the mountains where the rivers originate, this splitting will go on and on. There will be eventually quite tiny little streams feeding the river. Mathematically the structure forms what would be called a tree, for fairly obvious reasons. With each link in this tree, one can associate variables, for example, the catchment area of the section, the area of ground into which rainfall drains, the average slope, the average flow of water, and so on. That is one example of a tree. A tree in the botanical sense provides two examples of the tree because the part above the ground has this general structure and so do the roots. It was noted by Horton, an American engineer, fifty to seventy years ago, that even if one takes rivers from different parts of the world, there are remarkable empirical statistical regularities between these various variables. Similarly for trees, botanical trees, but they are different in detail, and there are other biological examples.

These structures are in some sense very probabilistic. What sort of probability model might have generated Horton's Laws? Secondly, from a more statistical point of view, what kinds of methods of statistical analysis might one use to try and capture both the essential similarities of these different structures, but also to discriminate between the different structures to show that the structure of the upper part of the river is different from a river delta where there is a somewhat similar structure going on.

What sort of a statistical design would come in if you started to consider the problem of where to put rain gauges or flow meters across this system in order to get the most efficient set of data?

Another apparently quite different, but somewhat related problem, occurs in phylogenetics: geneticists look backwards, through the evolution of a species or maybe of a micro-organism and construct a phylogeny, usually starting from an original individual repeatedly subject to binary fission. The branches of the binary tree that are extinct now will never be observed. So, given this kind of branching process observed looking backwards, what sort of probability model can be used? Does the model fit the data? How does one use the model to interpret features of the species? Again, a mixture of applied probability and statistical analysis specific to this is needed.

So far I have discussed statistics, not statisticians; the title did say "statistics". It is one of the aspects of statistics that is most encouraging from an intellectual point of view, but perhaps has its negative aspects from an individual point of view, that these ideas are, if relevant at all, likely to be sufficiently important to serious scientists working in these fields not to leave the job to statisticians. The scientists themselves will want to take some active part. That has been the case for a long time.

Applied probability in physics started with Maxwell and Boltzmann and the early work of Clausius on the kinetic theory of gases. It developed into classical statistical mechanics, a field rather mysterious to probabilists because of the rather unusual role of probability. More recently, however, there is some convergence between the study of statistical models in physics and the mainstream of probability. The role of statistical analysis in some areas of physics is also increasing, beyond direct uses of least squares and the propagation of error formula (the delta method).

CONCLUSION

This is a brief outline of why statistical ideas are of widespread importance in science. This is appreciated very explicitly in some areas and not so much in others. The job of the statistician, which I have not said much about, is to try to choose some areas of work out of the broad picture that I have illustrated, that one hopes will have scientific impact, perhaps immediately, but if not immediately, then certainly in the longer term.

CHAPTER 20

The Price of Ideas

A.T. Stewart

INTRODUCTION

I shall discuss here three facets of "The Price of Ideas". First, an example of a well-managed research laboratory, the source of ideas; secondly, the dangers of downsizing the research and development efforts in Canada; and finally, a philosophical argument which merges the concepts of "pure" and "applied" research and asserts the necessity of research for the survival of our society.

AN EXAMPLE

The purpose of research laboratories and institutes is to create ideas. Management of these organizations is not simple. An example of a well-run and very productive organization is the Chalk River Laboratories of Atomic Energy of Canada of the 1950s and 1960s. This is the best example I know of good management and good research leading to the production of great ideas.

The Chalk River Laboratories were directed, on the research side, by a remarkable man, W.B. Lewis. Lewis had been an electronics scientist during the war and came out to Canada to be research director at Chalk River when Sir John Cockcroft returned to England at the beginning of the Atomic Energy Research Establishment at Harwell.

I went to Chalk River in 1952 with a new Ph.D. from Cambridge. After the spartan but free life of a graduate student, the constraints of a research organization seemed somewhat onerous. Buses ensured a timely arrival at work; unless you missed them and had to catch a later one. But the most confining requirement was the Quarterly Progress Report. Every three months we had to write it, progress or not ! I recall that about a week after one such report, a memo arrived. Dr. Lewis noted that in my report I had written that a particular job was finished, had been written up, and had been submitted for publication. Lewis went on to suppose that in the next report I would say that the work had been published. "When," he asked, "are you going to tell me what it is you have actually done ?" So I did. There was no vacuum at the top when Lewis was there. He read everybody's report and often made comments and asked questions. He did this while he was directing a first-class team of engineers and scientists towards the design and engineering of power reactors. Lewis is the one person who knew every tree in the woods in detail and kept an Olympian view of the whole forest as well.

In other areas, too, the work at Chalk River was first class. In the Physics Division some nuclear physicists had surprised the world by showing that collective motions existed even in light nuclei. And the neutrons from the NRX reactor, one of the world's best at that time, were used for many practical experiments. Materials for future power reactors had to be tested in a high radiation field. These materials were placed in the reactor and the changes in strength, dimensions and many other structural characteristics of engineering design were measured.

The neutrons from the reactor also supplied B.N. Brockhouse and colleagues with the means to observe the motion of atoms in solids. This research certainly led the world, and for all his work Brockhouse was awarded the Nobel Prize in 1994. It was an exciting time.

This scientific research was hardly related to the design of reactors but it was supported by the management of W.B. Lewis. Today this research would be called "pure" or more obnoxiously "curiosity driven" research. Why was it being done at an establishment dedicated to reactor design?

The answer lies in the kind of management practised by Lewis. Some of these very good scientists were from time to time conscripted by Lewis to work with a team on a particularly difficult practical problem. When the problem was solved the scientist returned to his "purer" research. There were no barriers between divisions or departments or groups that hindered Lewis when he needed to assemble a team to work on a particular problem. Ignoring departmental labels and jurisdictional boundaries in order to solve specific problems is the best formula for success. It has been practised in many settings and always works if the leader is competent. Dr. Lewis was. In addition, he liked to have around him bright, informed, and imaginative scientists. The NRX reactor was a world attraction for such scientists and thus Lewis had his pool of talent which produced and tested ideas. It is a good system known to work well. It is the price of ideas.

FISCAL RESPONSIBILITY

There is no question that reductions in public expenditure are necessary if we are to leave our children with anything but a legacy of debt. The big problem is discrimination. When survival is at stake, what items in the Research and Development (R&D) sector can be considered luxuries? For survival alone, clearly much "pure" research is not needed; nor are ballet schools, or telescopes, or museums. We are not quite in that state but who has the wisdom and fortitude to set priorities? It takes a long time to recreate an orchestra after it has been abandoned, and it takes a longer time to recreate a science research group. But sometimes even the activities that make our society rich and worthwhile have to be dropped.

Those laboratories that have some relationship with industrial production or with health care require even wiser judgement. The support of many laboratories and institutions of this category is the minimum price of ideas and of society's survival.

An example of reductions in R&D is the reduction of staff in the Atomic Energy of Canada Limited at Chalk River as well as reductions in the staff of the R&D sections of Ontario Hydro that are related to the CANDU reactor system. Managing these reductions should be of great concern to the residents of Ontario since about two-thirds of the electricity supply comes from CANDU nuclear reactors.

A recent newspaper story about the Chalk River Laboratories stated that there would, in future, be no research that was not directly related to CANDU reactors. This policy is reasonable although it does not follow the W.B. Lewis philosophy of mixing "applied" and "pure" scientists. However, I am very concerned that it and its implementation may spell the end to long-term nuclear industry in Canada. There are three separate dangers: first, any industry with a large technological base must be continuously designing upgraded versions of its product. In this case there must be a series of improved CANDU reactors on the drawing board. And in addition, there must be a continuous stream of possible improvements and modifications to existing CANDU reactors. To achieve this state, a team of engineering and scientific staff must be kept together doing challenging work and, following the Lewis pattern, some basic scientists should be included too. While it is certain that officers of the Atomic Energy of Canada Ltd. (AECL) are trying to preserve the design team, the importance of the design team may be lost to those who define austerity as just barely keeping the present CANDU reactors maintained and running; and ageing, and ageing. That policy would be disastrous for a province that is presently dependent on nuclear reactors for most of its electrical power. Secondly, the implemention of cutting back is more likely to be done by abolishing a group or division or section than by a judgement of the contribution each individual might make to the overall CANDU program. Cancelling a program or abolishing a division is administratively easier to do, and the alternative requires a good knowledge of the individuals in each division. Abolishing by organizational units may result in damage to the CANDU project itself. It is an unthinking response to austerity. Thirdly, jurisdictional responsibilities are tangled. Ontario Hydro has a primary task of providing economic power to the province, and the Atomic Energy of Canada Ltd. research laboratories are primarily involved with the continuing R&D associated with the CANDU reactor. However, when the philosophy of "user pay for research" grows, and both organizations have severe budget reductions, the research itself suffers.

In this discussion of reducing costs, which is necessary, there is little public discussion of the need to build a new research reactor. The present research reactor at AECL, called NRU, has served very well for more than thirty years. It must be replaced soon, within the next few years. The present principal client or user of the NRU reactor is the electric utility industry at about $80 million worth of research and development per year. They need the facility! I hope that the leadership and staff that remain after the reductions and disgruntled resignations will be up to the challenges of the future. Is our society unwilling to pay the price for ideas so essential for the longer term future ?

PHILOSOPHY OF RESEARCH

In the previous discussion it is clear that the basic research that is connected to applications is of great value to our well-being. In this generation our health has improved, our clothing almost totally changed, our entertainment revolutionized, our communications and access to information unbelievably enhanced, and travel is so available that some say the whole world is a village. All of these changes have come about through new ideas being tested, changed, discarded or adopted. The Wright Brothers, or Graham Bell in Canada, did not call themselves researchers but their activities certainly were applied research. They and all the inventors of the past produced ideas and tested them. Should society support today's idea producers?

While it is easy to see what concrete benefits come from applied research and from basic research associated with possible applications, what is the point of doing research which seems to have no applications? Why do scientists pursue such problems?

There is a simple answer to the last question; unraveling the ways of nature is one of man's most fascinating occupations. The contest between man's mind and nature's secrets — the unrelenting quest to understand the hows, the whys, the whens, of the natural world — is for many the ideal life. The only sure reward of this quest is a clearer understanding of some small part of nature, but to many dedicated scientists this reward is sufficient.

In addition to the rewards to the scientist, society also is often rewarded, sometimes from unexpected sources. About fifty years ago streptomycin, the first effective treatment for tuberculosis, was discovered in a laboratory in an agricultural school in Rutgers University. This small and, until then, not well-known laboratory for soil microbiology was successful when the large professional institutes had failed. The lesson is that at the ideas level, small laboratories, small groups, or individuals pursuing basic research produce new and innovative ideas.

It is interesting to ask how long it would take a dedicated team of ophthalmologists to invent laser surgery, or a dentist to invent X-rays. The fast spread of knowledge and easy communication between disciplines has reaped great benefits for mankind.

. . . .

But the question, "What is the use of all this research?", has not been answered. We have cited only examples of some research that has produced useful results. We have not faced the basic question: Why pursue knowledge that is not likely to be useful? Before answering, let us consider the close connection that exists between research for pure knowledge and research aimed at practical results.

The principles of radio were demonstrated by Heinrich Hertz well before the turn of the century. But not until the Second World War were radio and radar developed intensively. After the war many scientists returned to university research and brought with them the modern (late 1940s) electronic techniques they had developed. This applied research helped the search for pure knowledge, a mutual dependence not always appreciated. One of the results of their new skills was the detection of the magnetism of some atomic nuclei. (Something less than one-half of all atomic nuclei behave like tiny magnets.) This advance in pure knowledge was of great interest to some scientists but at the time of little interest to anyone else. But now, nuclear magnetic resonance techniques are widely used in all chemistry laboratories, industrial and academic. Geologists use the technique for exploration and most large hospitals have a Magnetic Resonance Imaging machine which yields beautifully detailed pictures of organs in the body, all non-invasively.

This multiple interdependence of pure knowledge and knowledge that can be gainfully employed is characteristic of all man's understanding of nature. No one can say with certainty that some particular area of knowledge is useless to mankind, although admittedly it is easy to anticipate that some areas will not be used in the immediate future. Knowledge of nature is a whole knowledge and does not come in two categories, useful and non-useful. The utilitarian answer to the question, "Why research?", *is* that practical knowledge grows faster if a sufficient amount of pure knowledge grows with it, a philosophy practised by W.B. Lewis in the early days of the Chalk River Laboratories of Atomic Energy of Canada Ltd.

. . . .

There is an important philosophic answer, too. Aside from direct usefulness, research, in its broadest sense, is as necessary to mankind as food. In order to live and grow, man must struggle and must succeed or fail. The great efforts that send astronauts into orbits, and the scientist's search for true descriptions of nature's habits and laws, these are all necessary struggles. It is essential for all men that some individuals reach for the stars.

PART VII

CONCERNS OF SCIENTISTS IN PUBLIC POLICY

CHAPTER 21

A Statistical Excursion into the Arena of Public Policy

D.J. Bartholomew

INTRODUCTION AND BACKGROUND

The purpose of this paper is to describe and draw lessons from the experience of the Royal Statistical Society's Working Party on the Measurement of Unemployment in the United Kingdom.

The origins lay in a debate, extending over many years, about the validity of the United Kingdom unemployment figures. This debate came to a head early in 1994 when John Prescott, then shadow employment secretary, declared that the figures were "fiddled", and were no longer worthy of serious discussion.

In anticipation of a request to the society from Prescott to examine the matter, the council of the society set up a working party. This group consisted of the author as president, Peter Moore and Fred Smith, the two immediate past presidents, and Paul Allin, the honorary meetings secretary. Three basic principles underlay the group's manner of working: (i) it must be, and be seen to be, completely independent; (ii) it must work with the full knowledge and support of both Government and Opposition; and (iii) its recommendations must be framed in the light of what the users of the figures required. These aims were achieved and, in particular, the impartiality of the working party was never questioned.

The report was published in April 1995 and subsequently appeared in the *Journal of the Royal Statistical Society* (Bartholomew, Moore, Smith and Allin, 1995) together with discussions. Full details are given there, but the following outline sets the main substance of this note in context.

The monthly rate, in dispute, is based on the number of people claiming unemployment benefit, the Claimant Count (CC). This is produced speedily, accurately, and can be broken down to give information on local areas. The problem is that the rules governing the benefit have changed over thirty times since 1979 and almost all changes have resulted in a reduction in the total number of "unemployed". Changes in real unemployment were thus confounded with changes in definition. The effects were not large, but this flaw had badly dented the credibility of the index.

A second, quarterly, index based on the Labour Force Survey (LFS) is also produced. This measure uses an internationally accepted definition of unemployment. In the recent past the two measures had been very close, but discrepancies have occurred and were, we thought, likely to occur again. The main disadvantage of the LFS measure is that the sample is not large enough to allow small area comparisons.

We recommended that the Claimant Count should continue to be used, but that the LFS measure should replace it as the main headline figure. Our preference was for it to be produced monthly but, if this proved too costly, we suggested ways of deriving monthly figures from the LFS data which are collected continuously. In addition, we proposed several subsidiary measures, to focus on such things as youth and long-term unemployment.

PUBLIC RESPONSE

The report received substantial, supportive coverage by the media. The quality press and the BBC *Today* program, for example, treated the issues seriously and accurately. The response of both Government and Opposition was respectful but disappointing, in that both tended to reiterate their previously held positions, for which they now claimed our support! The views expressed in public by both sides were less subtle than those we had heard in our private discussions.

One of the most revealing experiences of the difficulties of communication between statisticians and politicians occurred when Peter Moore and I appeared before the House of Commons Select Committee on Employment. The committee was chaired by a Labour member and consisted of roughly equal numbers of Conservative and Labour

MPs. Its sessions were held in public and, in our case, excerpts were subsequently televised. The session lasted over two hours and immediately revealed a degree of mutual incomprehension. We were expected to be walking encyclopedias on the benefit rules, and to know all conceivably relevant figures by heart. The statistician's purpose was expected to be removal of uncertainty rather than its measurement. It was not understood why a sample survey was not necessarily worse than a complete count. The politicians were unable to distinguish between a statistical question and a political one.

Most disconcerting of all, perhaps, was the adversarial character of the questioning. We had supposed that, as expert witnesses, we had been called to provide professional help to the committee. Instead, we were treated as hostile witnesses whose case had to be discredited. Doubtless, this was mainly a matter of style, for which we should have been better prepared. We were, therefore, pleasurably surprised to find the committee's recommendations were very much in line with our own.

The Office of National Statistics subsequently set up a group headed by David Steel to examine the options for producing a monthly index based on the LFS. It provided a very thorough analysis of the various sampling and estimation methods, together with their costs.

At the time of writing (November 1996) it appears that the government has rejected the idea of using this as the monthly headline figure but has agreed to give greater prominence to the present quarterly LFS measure. This may not be the last word on the matter.

REFLECTIONS

The following reflections may be helpful to others embarking on a similar exercise. First, the statistician sees the social and economic world as inherently uncertain and the data are often fragmentary and inaccurate. This means that certainty is unattainable and the results must necessarily be couched in terms of probability. The politician, on the other hand, sees indecisiveness and uncertainty as marks of weakness. Unless the statistician can remove those uncertainties and provide a base for a confident decision, the statistician is of little use. Politicians, for their part, are elected to change things and believe that they have the power to do so. Statisticians know that control is a highly uncertain business. Politicians judge the rightness of an action retrospectively; statisticians judge the rightness of the logic which led to the action, in full recognition that even the best decision-making logic cannot deliver the desired result on every occasion.

CONCLUSION

This divide will not be easily overcome, but mutual recognition of the following two principles would be a modest first step.

First, modelling is an essential element in rational thought. We all have models, explicit or implicit, politicians especially. The only virtue that the statistician can claim is to make them explicit, and therefore open to testing and scrutiny. Once this fact is recognized by everyone, the agenda is set for a more fruitful discussion. And finally, there is a logic of uncertainty that imposes a discipline upon our reasoning. In the social context especially, we must live with the fact that no amount of analysis, however sophisticated, can wholly eliminate uncertainty which means that, sometimes, we shall be wrong.

REFERENCE

Bartholomew, D.J., Moore, P.G., Smith, T.M.F. and Allin, P. (1995), "The Measurement of Unemployment in the UK" (Report of a Working Party) (with discussion), *Journal of the Royal Statistical Society A* 158:363-417.

CHAPTER 22

Scientists and Responsibility in Public Policy Making

G.H. Reynolds

INTRODUCTION

The manner in which reliability and integrity in science are compromised has more to do with subtle and unconscious personal and institutional biases than with deliberate dishonesty. Yet, the most common critiques of science emphasize intentional misconduct in research, and solutions have focused on reprimanding individual scientists. Clearly, other aspects of the research process need attention as well. Rather than focusing on the actions of individual scientists, revamping the institutional setting where research is conducted may be a better place to focus corrective action. There is a considerable amount of information on scientific responsibility in conducting research. Scientists are expected to be bound by values, traditions, norms and stands that exemplify honesty, integrity, objectivity and collegiality (Broad and Wade 1982; Council of Biology Editors 1990; Panel on Scientific Responsibilities and the Conduct of Research, 1986). However, these principles and practices depend on scientists within scientific disciplines and on the self-correcting mechanisms inherent in the nature of science to ensure that scientists abide by these norms.

According to this model, objectivity is embedded in the social structure of science. Science is represented as a rigorous discipline that self-corrects for its biases, and scientists are part of a collegial and objective community of truth-seekers. However, other models of science may be helpful in recognizing and correcting biases that compromise the integrity of the scientific method (Reynolds 1993, 1996). Further, the self-correcting mechanisms of science may not be adequate, given the political, economic and social context in which science is sponsored, funded, created and reviewed. Unacknowledged and/or unrecognized bias is perhaps the most serious threat to the proper functioning of the scientific method. In this paper, science is treated as a product of specific social, cultural, political and/or economic structure, and/or norms that determine what questions are asked and how they are answered. Responsible scientists and concerned policy makers must be willing to scrutinize these norms and find ways to compensate for the systematic biases these norms bring to scientific observations. In my view, these unacknowledged biases are the most serious threat to the integrity of science and go largely unrecognized. Consequently, this paper will focus on the subject of bias and the responsibilities of policy makers and scientists in relation to our mutual goal of assuring public good in health and welfare.

Because policy makers depend heavily on scientific work, and because public policy has a tremendous effect on the conditions in which scientific work is produced, both scientists and policy makers need to be concerned about preserving the integrity of the scientific process at all stages. These stages include: (i) posing scientific questions, conducting research and interpreting results; (ii) peer reviewing and publication, scientific debate, criticism and reproduction/verification of results; and (iii) communicating scientific information within and outside the scientific community.

POSING SCIENTIFIC QUESTIONS, CONDUCTING RESEARCH AND INTERPRETING RESULTS

The directions that science takes are determined by the economic and political contexts in which science is produced. Science occurs in a number of cultural sites, including but not limited to: universities, industry and government, all funded and sponsored for specific reasons that have a specific effect on what types of phenomena are being studied; how hypotheses are formulated; and how conclusions are drawn. The interests that underlie a particular piece of research and its results can affect the way scientific questions are posed, which questions get posed and which do not, whether there is healthy debate, and whether policy makers are receptive to the results (Reynolds 1993, 1996).

Economic Interests

Historically, universities have carried out basic research with an "appropriate disinterest" in the results. However, with increasing ideological and fiscal attacks on higher education, the future growth of university research will depend increasingly on corporate sponsorship, a development that has been encouraged as part of a national science policy.

Cultural Bias

In examining what questions are posed, one should also consider the subtleties of cultural bias; these questions are often situated in common-sense notions that come from cultural ideologies, which in turn will situate the scientific investigations. These common-sense notions conflict with the scientific method by acting as invisible premises (ibid.). The biases instilled by our culture (xenophobia, homophobia, racism, sexism, classism) become situated in our scientific investigations and our public policy. The virtues of liberal individualism (competition, personal responsibility, self-reliance) and of capitalism (profit, privatization, self-interest, political influence) appear in scientific research as much as they do in U.S. history and public policies. Examples of these values in public health are the studies that emphasize the behaviours of individuals as causes for poverty, criminal activity and low educational achievement, rather than looking at broader structural, economic or cultural causes (Freeman, 1993; United States Department of Health and Human Services, 1985; Reynolds, 1996; Waitzkin, 1983). Such studies encourage the notion that such social problems are caused by the moral failings of individuals and resist any analysis of the context in which these problems occur. Hence, we see a number of simplistic genetic explanations offered for complex social phenomena. Specific genes are sought for intelligence, for obesity, for homosexuality, for violence and for mental illness which policy makers are very willing to support (Horgan, 1995; Reynolds, 1996). When these genes are "found", scientists, policy makers, and the media report with great fanfare and hype. Behaviour, these findings imply, can be completely understood at the individual level. And when many of these studies either cannot be replicated or are later retracted, the media basically ignores the retraction and the public and policy makers rarely if ever hear of the retractions. Thus, the cultural biases of the public and policy makers are reinforced.

Misconduct and Bias

Scientific misconduct is defined as "fabrication, falsification, or plagiarism, in proposing, performing, or reporting research" (Panel on Scientific Responsibility and the Conduct of Research, 1986). When such practices come to light, they erode the trust that policy makers and the public have in scientists and their results (Broad and Wade, 1982; Bross, 1987; Crossen, 1993; Panel on Scientific Responsibility and the Conduct of Research, 1986). Much has also been written about other factors such as questionable research practices that are not considered misconduct, and these may be much more damaging to the integrity of the scientific process than scientific misconduct. These factors include: inappropriate design or methodology; errors in judgement; errors in selecting, analyzing or interpreting data and in using inappropriate methods of measurement to enhance the significance of research findings; inadequately supervising research subordinates or even exploiting them; misrepresenting speculations as fact; and releasing preliminary research results, especially to the public, without providing sufficient data to allow peers to judge the validity of the results or to reproduce the results (Bailar, 1985*a*,*b*, 1986; Broad and Wade, 1982; Bross, 1987; Crossen, 1993; Grinnell, 1992; McNeill, 1993).

There is much less discussion about the social and cultural subjectivity of our observations occurring in all scientific and non-scientific fields, and how the work is dependent on conscious and unconscious agendas on values (Guinan, 1993; Reynolds, 1993, 1996). Values and observations are both framed in a context of cultural and historical meanings and social forces. Scientists use their personal experience to decide what to observe, how to make and interpret observations, how to decide which questions to ask and how to interpret the results (Reynolds, 1993, 1996). Thus, both scientists and non-scientists alike bring issues of blame, of "innocent and not innocent" victims, or "worthy and not worthy" victims, to the interpretation of sickness and health, wealth and poverty (Herman and Chomsky, 1988; Reynolds, 1993). We learn from and rely on cultural narratives that are derived from the media and popular culture, and we use these to draw conclusions and create policy.

PEER REVIEW AND PUBLICATION, DEBATE/CRITICISM AND REPRODUCTION/VERIFICATION OF RESULTS

The peer-review process is designed to deal with many of the problems that could arise due to conflict of interest or other biases of individual researchers (Bailar, 1985*a*). This process, however, is vulnerable to the same problem of bias that it intends to resolve, that is, focusing more on the institutional setting in which the peer-review process functions. Again the integrity of the peer-review process depends on a diversity of viewpoints, an interdisciplinary approach, and well-informed researchers with independent motives.

A number of questions need to be asked about this process. First, when reviewing and approving research protocols and in the peer review of journal articles, is there a policy that qualified statisticians are included in the review? Is there too much emphasis on succinctness by editors and reviewers of papers which may be partly responsible for the omission of assumptions, the reporting of a study in multiple fragments, the reporting of only consistent and positive results? Is the pressure to publish, to be first, such that it encourages cooking, selecting, mishandling or misrepresenting data and does this interfere with the scientist's ability to think creatively, work rigorously or make contributions to the field? Does the peer-review system take adequate measures to check against conflict of interest scenarios? Scientists are often in the position of critiquing methods that they have either a professional or financial investment in. This has to affect their ability to be objective.

The following example highlights some of these problems: a number of papers evaluating a new diagnostic test for a sexually transmitted disease, *Chlamydia Trachomatis,* all used some version of a method of analysis called discrepant analysis. This "statistical method" clearly is biased and overestimates both specificity and sensitivity. Thus, tests which are evaluated using discrepant analysis present upwardly biased estimates of sensitivity and specificity. A large number of false positives and negatives are especially likely when the prevalence of the disease is fairly low (McGee and Reynolds, 1996). Obviously, presenting the test as more accurate than it is has detrimental practical and social consequences for those who use and take the test, but has a positive value for those who manufacture and sell the test (Danielsson, Falk and Persson, 1996).

However, papers using some version of discrepant analysis made it through the peer-review process and have been published by leading journals, including the *Journal of Clinical Microbiology, Sexually Transmitted Diseases, Genitourinary Medicine* and *The Lancet* (Hadgu, 1996*a,b*). The editors, reviewers and the international review boards all failed to understand the biases introduced by the use of discrepant analysis. Some still defend the methodology. One government statistician at the Centers for Disease Control and Prevention observed this problem and called it to the attention of the Food and Drug Administration. Subsequently, one company that manufactures a Chlamydia test was asked to change their claims of sensitivity and specificity in the package inserts. If this method had been allowed to be used in evaluating the Chlamydia tests, it would have set a precedent for approving many other tests for other diseases such as gonorrhea, herpes simplex virus, tuberculosis, prostate cancer, heart disease and testing for drug use. A policy blunder was avoided, but not by the normal peer-review process. A test may be useful for certain purposes, such as screening to determine who should have further testing with a more specific and sensitive test that may be more invasive or costly. But the health provider needs a fairly accurate estimate of sensitivity and specificity with which to begin.

Attempts by psychologists and physicians to measure intelligence earlier in this century provide an example of bias that has gone unchecked and continues to have appalling consequences. Scientists had long assumed that races could be ranked on an evolutionary ladder, with white non-Hispanics at the top, and that this was reflected in differences in the physical attributes of humans (Cronkhite, 1991; Crossen, 1993; Gould, 1981; Lewontin, Rose and Kamin, 1984; Tucker, 1994). Leading scientists of the day, in their "measurements" of intelligence, confirmed the theory, initially, with measurements of the skull, and later with IQ tests (Gould, 1981; Lewontin *et al.*, 1984; Tucker, 1994). Sam Morton and other leading scientists of the day believed that the volume of the skull is a measure of intelligence, and, of course, whites were on top. When Morton died, the South's leading medical journal said, "We of the South should consider him as our benefactor for aiding more materially in giving to the Negro his true position as an inferior race" (Broad and Wade, 1982). Paul Boca in France followed his lead. H.H. Goddard developed a scale of mental deficiency and said "that the feebleminded as identified by Goddard should not be allowed to breed lest they reduce further the average intelligence of the population." Goddard found among immigrants that 83 per cent of Jews, 80 per cent of Hungarians, 79 per cent of Italians and 87 per cent of Russians were feebleminded by his test. These early intelligence tests, used to "prove" that Jews, Hungarians, Italians, Russians and blacks were feeble-minded, were intended to help identify and curtail the reproduction of bad genes (Gould, 1981; Lewontin *et al.*, 1984; Tucker, 1994). The Eugenics Movement, for example, was initially a recognized and respected branch of genetics. Galton wrote that "If the poorer stock continued to procreate children, inferior in moral, intellectual and physical qualities, the time may come when such persons would be considered as enemies to the states." He and other eugeneticists published extensively in the leading journals of the time. Eventually this movement led to the practice of mass murder (Goldhagen, 1996; Pollack, 1994). Lewis Terman, Robert Yerkes and Cycil Burt, and others, we now know, believed so strongly in the idea of racial superiority that they fabricated data, deleted or selected data, misused statistics and misconstrued/misinterpreted the results (Broad and Wade, 1982; Cronkhite, 1991; Crossen, 1993; Gould, 1981). These absurd and atrocious results were used to justify many injustices with especially disastrous and long lasting influence. They include the testing and sorting of children at the age of eleven for many years, in Great Britain, resulting in 80 per cent of these children being called unfit for higher education because of low innate intellectual ability. In one state, Virginia, 7,500 women were sterilized because they were considered feeble-minded and/or anti-social between 1924 and 1972 (Gould, 1981). Thirty states had similar laws. The Virginia law which allowed the state to perform these sterilizations was upheld in a Supreme Court decision delivered by Oliver Wendell Holmes in 1927. Scientists supported eugenic reform of immigration laws and testified about the "latest evidence for the decline in genetic quality of immigrants". The mistaken notions about the mental inferiority of immigrant races, blacks and lower class whites were used in the arguments given for passing the *Immigration Restriction Act* of 1924 (Cronkhite, 1991; Gould, 1981; Tucker, 1994). As a result thousands of Jewish refugees were not admitted into the U.S. during the 1930s.

These were not isolated inquiries. Throughout the 19th and 20th centuries some scientists searched for some measures of man that would rank some groups higher than others. These scientists claimed to be strictly objective and unbiased and always these rankings came up with justification of the social prejudices of the time and were used to justify injustices such as ethnic cleansing, slavery, segregation, subjugation of women or European domination over other people. Paul McNeill states in his book, *The Ethics and Politics of Human Experimentation*, that "The history of unethical experimentation is an appalling account of man's inhumanity to man" (McNeill, 1993). These types of experiments were reported in scientific and medical journals and again were well known to scientists. Respected medical researchers tortured slaves and helpless Irish indigent women in their experimental work on which modern gynecological surgery is based. Medical doctors and other scientists in Germany and Japan performed the most abhorrent experimentation on prisoners and civilians before and during WWII (Goldhagen, 1996; McNeill, 1993; Pollack, 1994). In Australia, experiments using mustard gas on human subjects were carried out. The British government conducted nuclear testing on human subjects in Australia and showed little regard for the safety of its citizens, especially the Australian Aborigines. In the United States a series of experiments that lasted until the early 1970s tested the effects of nuclear radiation on humans and were intended to cause harm to the subjects and in other countries there are still reports on physician complicity in torture (Iacopino, Heisler, Pishevar, and Kirschner, 1996; McNeill, 1993; Westmeyer, 1996). Psychometric testing had great scientific prestige among informed psychological opinion and has had continued support throughout this century, despite many demonstrations that the premises of such tests have no scientific validity (Gould, 1981; Tucker, 1994). IQ tests, SATs and similar standardized tests are still heavily relied on to determine (or rather, justify) who is given access to higher education. William H. Tucker in *The Science and Politics of Racial Research* recounts how scientific research has been misused to justify racism and oppressive political positions, for example, the work of scientists such as Galton, East, Cattell, Jensen, Shockley, etc., and *vice versa*, the support by think tanks of this type of research and its use to make public policy (Tucker, 1994).

The lack of diverse viewpoints in science and public policy in terms of race, ethnicity, class and social backgrounds, is a major concern, especially for those working in the area of minority health; an area significantly affected by the cultural biases of scientists and public policy makers (ACE, 1995; Greenberg, 1993; Greenberg, Buffler, Himman, Swanson, Matanoski, Nasca and Bracken, 1995; Guinan, 1993; Omi and Winant, 1989; Pearce, 1996; Reynolds, 1993, 1996; Schoenbach, Reynolds and Kumanyika, 1994; Warren, 1993). Minority health is impacted by many scientific disciplines (economics, anthropology, epidemiology, medicine, sociology, engineering, biology, criminology, psychology, statistics) and by many areas of public policy (education, environment, public health, criminal justice, assistance programs, rehabilitation) which persistently embrace current prejudices. Severe disparities in health outcomes exist between people of colour and white populations in the U.S. (Cooper, 1993; Freeman, 1993; National Center for Health Statistics, 1995; United States Department of Health and Human Services, 1985). Unless the solutions are framed in the historical, political, economic, legal and social context, we have little chance of closing the gap in these health inequities.

THE COMMUNICATION OF SCIENTIFIC INFORMATION WITHIN AND OUTSIDE THE SCIENTIFIC COMMUNITY

According to Howard Zinn (Zinn, 1990, 1995), "Society has varying and conflicting interests; what is called objectivity is the disguise of one of these interests — that of neutrality. But neutrality is a fiction in an unneutal world." It is probably more often a myth than not and perhaps unrealistic to expect that scientists can always be objective truth-seekers. This may be analogous to the myth of objective truth-seeking journalism in modern, capitalist, "democratic" nations. As Noam Chomsky and others have demonstrated, the "free press" is free insomuch as it is not subject to totalitarian control, but no more (Chomsky, 1996; Herman and Chomsky, 1988). The mass media are a capitalist enterprise, not a public service and the absence of dissent is explainable by this fact alone. However, the fact that most journalism is a product created for profit and paid for by corporations should make it clear that the media and the journalists who often come from similar historical, political, economic, legal and social backgrounds, will reinforce and even magnify biases when they bring scientific information to the public or discuss policy decisions based on scientific findings.

Scientists have a responsibility both to the public and to makers of policy to communicate scientific information in a way that explains the processes and protocols used, the limitations of the study, and why scientists come to those conclusions. Scientists need to acknowledge the cultural and social assumptions and implications of their research and need to ensure that there is continuing open discussion between the different disciplines and the public. They need to individually and collectively ensure that the results are used appropriately in all settings and that our science serves the public good. Scientists who wish to remain neutral with regard to societal and political uses and misuses of their research have made a choice. Not making a choice is making a choice!

CONCLUSION

Scientists need to rethink their responsibilities as scientists in setting the agenda for public policy by focusing on societal needs. If we desire scientific, social and public policy influence we also need to take social responsibility for the use and misuse of our work. Lack of concern about the consequences of our research is abdicating our responsibility as citizens and as members of the human race (Zinn 1990, 1995). It will do little to promote integrity in science if we focus only on the actions of scientists as individual actors; in order to minimize cultural prejudices and other biases, we must focus on the institutional changes that are needed; we must set standards that guide the research and policy-making processes; and we must create a more culturally diverse and interdisciplinary environment for conducting our scientific work. Central to this effort is a commitment to both recognizing and countering the pervasive bias that exists in both science and public policy.

The scientific process ensures that the work of individual scientists is evaluated by their peers; meaning and significance is given to scientific conclusions because they represent the consensus of experts in a number of fields. Ideally, this process corrects for any bias that a particular scientist might have; it is a system of checks and balances. Since no such process exists for the creation of public policy, there are no such checks and balances. Statisticians and other scientists must take more responsibility to ensure that the right questions and objectives are posed and that the results are correctly interpreted and used.

Specificity, it is our responsibility, as scientists and policy makers, to:

(i) *Deal with the historical legacies*, that is, the cultural and historical origins, meanings and forces that have affected and will continue to influence every aspect of life and public policy, whether it is in education, employment, the environment, criminal and civic justice, assistance in transfer programs or health-care access.

(ii) *Broaden the objectives and scope of both scientific and public policy debates*, for example, the relationships between public health, social justice, racism, social class, poverty, access to health care, equitable distribution of wealth, health and quality of life.

(iii) *Increase communication and interdisciplinary co-operation among scientists, policy makers, the community and the media.* This will encourage not only a more accurate representation of the results but of the purposes of the research itself — that is, who sponsored the research and why, whether further research is needed, what are the underlying assumptions, and most important, ensure that the results are not misused by special interest groups, policy makers, etc.

(iv) *Insist that scientific methodology is sound* and that the necessary studies and monitoring or quality control systems are implemented to safeguard against errors which can lead to subsequent mistakes and injustices in health and public policy.

(v) *Utilize our professional organizations* to foster inclusiveness and to bring racial, ethnic and cultural diversity to their fields.

(vi) *Insist that we as scientists and our professional organizations do not hide behind the cloak of neutrality* to shirk responsibility for the type of research that is sponsored and funded, for the interpretation of results and for the uses and misuses of research.

REFERENCES

ACE (1995), Statement of Principles, *Annals of Epidemiology* 5:508.

Bailar, J.C. (1985*a*), "Journal Peer Review: The Need for a Research Agenda", *The New England Journal of Medicine* 312(10):654-57.

________ (1985*b*), "When Research Results Are in Conflict", *The New England Journal of Medicine* 313(17):1080-81

________ (1986), "Science, Statistics, and Deception", *Annals of Internal Medicine* 104:259-60.

Broad, W. and Wade, N. (1982), *Betrayers of the Truth: Fraud and Deceit in the Halls of Science*, New York: Simon and Schuster.

Bross, I.D. (1987), *Crimes of Official Science: A Casebook*, New York: Biomedical Metatechnology Press.

Chomsky, N. (1996), *World Orders: Old and New*, New York: Columbia University Press.

Cooper, R.S. (1993), "Health and the Social Status of Blacks in the United States", *Annals of Epidemiology* 3(2):137-44.

Council of Biology Editors, Editorial Policy Committee (1990), *Ethics and Policy in Scientific Publication*, Council of Biology Editors, Inc.

Cronkhite, J. (1991), "Guesswork — The Pseudoscience Behind the SAT," *The North Avenue Review*, June.

Crossen, C. (1993), *Tainted Truth*, New York: Simon and Schuster.

Danielsson, D., Falk, L. and Persson, K. (1996), "Discrepant Analysis and Screening of Chlamydia Trachomatis; Letter to the Editor", *The Lancet* 348:1308.

Freeman, H.P. (1993), "Poverty, Race, Racism, and Survival", *Annals of Epidemiology* 3(2):145-49.

Goldhagen, D.J. (1996), *Hitler's Willing Executioners*, New York: Knopf.

Gould, S.J. (1981), *The Mismeasure of Man*, New York: W.W. Norton.

Greenberg, R.S. (1993), "Editorial: Is Epidemiology Broken Down by Race and Ethnicity?", *Annals of Epidemiology* 4:337.

Greenberg, R.S., Buffler, P.A., Hinman, A.R., Swanson, G.M., Matanoski, G.M., Nasca, P.C. and Bracken, M.B. (1995), "Commentary — American College of Epidemiology Statement of Principles", *Annals of Epidemiology* 6:503-04.

Grinnell, F. (1992), *The Scientific Attitude*, New York: The Guilford Press.

Guinan, M.E. (1993), "The Myth of Science as a Value-Free Discipline", New York: Scholarly Publishing, pp. 263-68.

Hadgu, A. (1996*a*), "Discrepant Analysis and Screening of Chlamydia Trachomatis; Letter to the Editor", *The Lancet* 348:1309.

________ (1996*b*), "The Discrepancy in Discrepant Analysis", *The Lancet* 348:592-93.

Herman, E.S., and Chomsky, N. (1988), *Manufacturing Consent: The Political Economy of the Mass Media*, New York: Pantheon Books.

Horgan, J. (1995), "Gay Genes, Revisited", *Scientific American* 273:26.

Iacopino, V., Heisler, M., Pishevar, S. and Kirschner (1996), "Physician Complicity in Misrepresentation and Omission of Evidence of Torture in Postdetention Medical Examinations in Turkey", *Journal of the American Medical Association* 276(5):396-402.

Lewontin, R.C., Rose, S. and Kamin, L. (1984), *Not in Our Genes*, New York: Pantheon Books.

McGee, D.L. and Reynolds, G.H. (1996), "Discrepant Analysis and Screening of Chlamydia Trachomatis; Letter to the Editor," *The Lancet* 348:1307-08.

McNeill, P.M. (1993), *The Ethics and Politics of Human Experimentation*, Cambridge: Cambridge University Press.

National Center for Health Statistics (1995), *Health, United States*, Hyattsville, MD: Public Health Service.

Omi, M. and Winant, H. (1989), *Racial Formation in the United States*, New York: Routledge.

Panel on Scientific Responsibility and the Conduct of Research (1986), *Responsible Science: The Impact of Technology on Science*, volumes 1 and 2, San Francisco: Harper & Row.

Pearce, N. (1996), "Traditional Epidemiology, Modern Epidemiology, and Public Health", *American Journal of Public Health* 86(5):68-83.

Pollack, R. (1994), *Signs of Life: The Language and Meanings of DNA*, Boston: Houghton Mifflin Company.

Reynolds, G.H. (1993), "Foreword", *Annals of Epidemiology* 3:119.

________ (1996), "Culture, Science and Public Policy", *Statistics in Epidemiology Report* 3(1).

Schoenbach, V.J., Reynolds, G.H. and Kumanyika, S.K. (1994), "Racial and Ethnic Distribution of Faculty, Students, and Fellows in US Epidemiology Degree Programs, 1992", *Annals of Epidemiology* 4: 259-65.

Tucker, W.H. (1994), *The Science and Politics of Racial Research*, Urbana and Chicago: University of Illinois Press.

United States. Department of Health and Human Services (1985), "Report of the Secretary's Task Force on Black and Minority Health", Washington, DC: Government Printing Office.

Waitzkin, H. (1983), *The Second Sickness: Contradictions of Capitalist Health Care*, New York: The Free Press.

Warren, R.C. (1993), "The Morbidity/Mortality Gap: What is the Problem?", *Annals of Epidemiology* 3:127-29.

Westmeyer, J. (1996), "Compromise, Complicity, and Torture", *Journal of the American Medical Association* 276:416-17.

Zinn, H. (1990), *The Politics of History*, Urbana and Chicago: University of Illinois.

________ (1995), *A People's History of the United States*, New York: Harper Perennial.

PART VIII

BANQUET SPEECH

CHAPTER 23

Government View of Science and Technology

P. Milliken

The discussion of the conference has focused on three themes and I should note that the order is interesting. Statistics is first, science is second and public policy is last. However, the politician, along with scientists and statisticians, help shape public policy, and indeed, the three subjects are closely meshed. We all have a stake in the outcome of scientific research. The quality of our lives has been vastly improved by this work and will continue to be. What is at issue is the interplay between government and scientists in determining how tax revenues are to be spent on scientific research.

As you know, the Canadian government recently went through a public consultation process on this issue and as a result has issued its new science strategy entitled *Science and Technology for the New Century*. This document, which makes clear that if we are to optimize returns from scientific work, states the government must concentrate its research efforts on its core activities and that these are to be funding of scientific research in order to support the mandates of the private sector research and development. And, in addition to these traditional activities, "the government will increasingly emphasize a new role: that of information analyst, knowledge disseminator and network builder — critical elements in the successful evolution of the Canadian innovation system."

The government also announced that it would introduce a new investment program to stimulate increased private sector investment in high technology product and process development. This will be a partnership arrangement and will be targeted in areas including aerospace and defence, environmental and enabling technologies and industries converting from military to civilian production.

Every Minister with a Science and Technology responsibility released a response to the strategy, and in the case of the Minister of Industry, the response covered his entire portfolio of departments and agencies, including the National Research Council.

The science policy was followed up recently by a new vision document issued by the National Research Council entitled *Vision to 2001* wherein the objectives for the National Research Council are set out. Among other things, the National Research Council states:

> We will determine how much of an impact the National Research Council has on Canadian firms and, thereby, on Canada's economy. Wealth-creation impacts including success in boosting industrial competitiveness and investments in R&D and technology in Canada will be monitored.... Our success in commercializing our technologies will be measured by the success of our industry partners using our technologies and by the other transfers of technology we accomplish. We will also report on the progress achieved in building an entrepreneurial organization through more effective management practices.

While some will find the fearsome hand of government control pervading all this, I do not see anything as sinister as some suggest. There will be criticism of the policy, of course. But that is the nature of politics and we must all strive to make policy and to improve it. Policies are never written in stone, as everyone here knows.

What is also true is that policy, science and statistics are tied together. Adam Holbrook of Simon Fraser University once wrote that the ability to apply technology to the pursuit of national economic objectives (or even social objectives) makes the understanding of technology an absolute necessity for decision makers. Yet, national decision makers frequently have little understanding of the technological resources at their disposal, and often rely on out-moded anecdotal evidence to support their technological decisions.

If you need evidence of the importance of technology in decision making, you need look no further than the Canadian government's budget cuts, driven as they are by the depressing economic figures on Canada's deficit situation and the projections that are made from them. There are a number of thoughts on government policy: first, politicians love to quote scientists as authorities. Why? Because scientists have a lot of credibility with the public. As already pointed out by Dr. Conquest, a speech by a scientist can still attract a crowd. In Kingston, a good scientist will attract a larger crowd than an all-candidates meeting held during an election campaign. I suspect the same is true elsewhere. Scientists to politicians are like scripture to priests. Do not, therefore, underestimate your importance to politicians.

Scientists can also bolster arguments. The recent cuts imposed by the Minister of Fisheries in Canada on the west coast were defended most vigorously by the Minister by using the latest scientific data. What other justification could he use? Even those defences brought howls of protest. Remember, politicians and journalists are ranked lower than lawyers in terms of trustworthiness in public opinion polls. Scientists are much higher up.

Secondly, scientists must advance their own interests and to do so inform the public of their activities and their knowledge. There have been excellent presentations on the ways and means of dissemination of information relating to research, and I would like to mention the importance of meeting with your local representative. How many of you have invited your Member of Parliament or Congressman to visit your workplace? Here are electors, not just the scientist, colleagues, researchers, lab technicians and secretaries. There is something interesting to learn that will also be important. And when the politician is faced with applications for funding, it is easier to understand research that has been seen.

Politicians also like photo opportunities for use in their published materials. A snapshot with an expert on a matter of public interest makes good copy. I know of the concern of many scientists about appearing to be seeking public recognition for their work when they make public appearances. This attitude needs to be corrected. It happens in every field of endeavour. Some politicians are media hogs too. But the public will not support expenditure of its money if the results of the expenditure are invisible. Visibility and relevance are important.

Thirdly, beware the tendency to use highly technical language in explaining your position to the public, including journalists and politicians. We are both easily lost. I will, however, quote from a paper presented by Bill Leiss to a Royal Society of Canada Symposium in June 1994 that while not strictly relevant in this context, conveys the sense of what I mean.

> I do not believe that university researchers are providing nearly enough support to citizens. And, I believe, the chief reason for this deficiency is the antiquated, indeed obsolete division of labour that is crystallized in the prevailing academic disciplines, in particular the arcane terminologies that exist primarily to protect the territorial interests of the practitioners in those disciplines. Finally, I believe that if this situation persists unchanged, university research will garner a diminishing share of public support for knowledge generation, as new institutional structures arise to meet the public demand for policy-relevant knowledge that is expressed in languages that citizens can readily comprehend.

Fourthly, get to know officials who are drafting and implementing policy. They will appreciate your expertise and advice, and it is often they who make policy, rubber stamping is all that is expected from the Cabinet.

The interplay between science and statistics is essential to public-policy formulation. You have become partners in our political life whether you like it or not since your data provide the basis for public discussion. Your expertise is needed and indeed is sought.

PART IX

SCIENCE AND THE PUBLIC PURSE

CHAPTER 24

Caesar and the Savants: Some Socio-Political Contexts of Science and Technology in Canada

J. Meisel

BACKGROUND

The history of science and technology, henceforth S&T, reveals a long parade of discoveries and applications driven in part by the curiosity of scientists and in part by the agendas of religious, dynastic, patriotic, commercial and other interests. At present, throughout the world, countries are hell-bent on pursuing, through S&T, the flickering grail of wealth, economic competitiveness, and something ambiguously called the "quality of life", defined invariably in terms of purely physical well-being. Canada has not remained aloof from this quest and S&T have consequently attracted renewed interest from policy makers.

Thus, in the early 1990s, the Liberal Party's famous *Red Book,* its heavily touted electoral platform in the contest which brought it to power (Liberal Party, 1993), placed its meagre coverage of S&T in a chapter entitled "An Innovative Economy". Although intellectual and scientific matters received only feather-weight attention in the party program, the 1993 Chrétien government did, in mid-1994, launch a major review of S&T.

In this paper I examine this exercise and its fruit[1] in two distinct contexts: that relating it to the nature of Canadian society and in the comparative perspective of British experience. I conclude by speculating about why Canada's S&T record, and the review process, are disappointing and about what might be done about it.

Before turning to the reports produced by the review process, it is necessary, however, to sketch the background: What was the S&T situation in Canada in the early 1990s? Like so many other things Canadian, this question presents a paradox. On the positive side, the country has created some very impressive government laboratories, university departments and research centres which have provided congenial settings for first-rate scientific work. Canada ranked sixth in the world in terms of the contribution of its scientists to the world's total number of publications in science, engineering and medicine, from 1981 to 1994 (May, 1997, p. 793). Canada always places among the top ten nations in the Scientific Citations Index (Friesen, 1997). When countries are ranked according to the number of papers cited in the Index, *weighted by population size*, Canada shares fourth and fifth place with Denmark, ahead of all G7 countries, but behind Switzerland, Israel, and Sweden (May, 1997, p. 794). On the applied side, Canada is a leader in several domains: agricultural technology, remote sensing, nuclear energy, space robotics, telecommunications and computer software, among others.

These bright spots are, however, overshadowed by some threatening features: a very high level of scientific illiteracy, an insufficient number of young scientists being trained for the tasks at hand, serious haemorrhaging of scientific talent to other countries, and profound financial crises shaking both public and private scientific institutions. Underlying these and other threats to the health of science in Canada are two deeply worrisome factors: massive declines in government expenditures on S&T, and particularly on pure research, and a flawed configuration of governmental institutions related to the well-being and development of science.

Among the tell-tale indicators of the seriousness with which a country seeks to benefit from science is the proportion of its resources allocated to research and development (R&D). Here, Canada ranked only fourteenth among the Organization for Economic Cooperation and Development countries in 1991. Table 1 shows that the Canadian gross domestic expenditures on R&D, as a percentage of gross domestic product (GDP) was 1.50 per cent in 1991, whereas the three leading countries (Sweden, Japan and Switzerland) chalked up the much more impressive percentages of 2.90, 2.87 and 2.86, respectively. The percentage rose to 1.61 per cent for 1993 and 1994 but declined to 1.59 in 1995

Table 1: International R&D Spending 1991

	*GERD U.S. $ M**	*GERD/GDP Percent*
Sweden	4,186	2.90
Japan	67,349	2.87
Switzerland	3,849	2.86
United States	154,348	2.75
Germany	35,563	2.66
France	25,033	2.42
United Kingdom	18,735	2.08
Finland	1,617	2.02
Netherlands	4,750	1.91
Norway	1,315	1.84
Denmark	1,535	1.69
Belgium	2,752	1.69
Austria	2,043	1.51
Canada	7,783	1.50
Australia	3,671	1.34
Italy	12,899	1.32
Ireland	420	1.04
Iceland	420	1.04
New Zealand	399	0.88
Spain	4,337	0.87
Portugal	502	0.61
Turkey	884	0.47
Greece	369	0.46
Luxembourg	**	**

Notes: GER – Gross Expenditures on Research and Development
GDP – Gross Domestic Product
OECD – Organization for Economic Co-operation and Development

* adjusted by OECD to obtain what it calls Purchasing Power Parity (PPP).
** data not available.

Source: Resource Book for Science and Technology Consultations – Industry Canada (1994).
Adapted from 1994 Report of the Auditor General of Canada.

(OECD, 1997). While the Government of Canada has been cutting funding for basic research, other countries, including the United States, the United Kingdom and France, have been increasing science allocations in their budgets, some by up to 40 per cent (Friesen, 1997, p. 2).

Total government expenditures on S&T were $7 billion in the early 1990s of which about one billion consisted of tax credits. The latter was reduced to about $650 million and the former to $5.7 billion for 1996-97. The total expenditure is slated to drop to $5.1 billion in 1997-98.[2]

By the end of 1997, it was estimated that about eight hundred scientists will have been let go by the Ottawa government alone. The world-famous Tandem Accelerator Superconducting Cyclotron at Chalk River, for example, was abolished by its owner, the publicly-owned Atomic Energy of Canada Ltd. This not only resulted in the destruction of an unique facility but also in the dispersal of highly qualified scientists, many of whom have had to leave Canada.[3] Likewise, the heretofore outstanding Geological Survey of Canada has had to make cutbacks which, *inter alia*, terminated fifty years of research on meteorite craters and led to the sacking of Alan Hildebrand, one of the world's top experts in impact-crater geology (Anderson, 1997*a*, *b*). Several years of salary freezes in the National Research Council have led to a unprecedented exodus of scientists to the private sector and abroad (McIntosh, 1997).

Private industry has until quite recently seriously underperformed in girding its work with research. A survey by the World Economic Forum placed Canada fourteenth among forty-five countries, ranked by the amount of money private firms allocated to S&T; the amount Canada spent was less than half of the OECD average (Industry Canada, 1996*b*; Auditor General, 1994, vol. 6, ch. 4). This state of affairs is, however, undergoing some change, as we shall see.

Despite the successes of Canadian science already noted, the scientific community and governments have been concerned about the country's performance in this area for quite a long time. Table 2, which lists the major initiatives and reviews of science policy since the middle 1960s, portrays a meandering and unfocused attempt to find the right path and a highly inconsistent series of measures and countermeasures seeking to enhance the country's management of research.[4] There were, as the table shows, more than ten inquiries in thirty years into the state of affairs and innumerable structural innovations intended to

Table 2: Major Science and Technology Reviews and Policy Initiatives, 1963-1994

1963	Report by Royal Commission on Government Organization (Glassco)
1964	Creation of Science Secretariat in the Privy Council Office
1966	Creation of Science Council of Canada
1968	Science Council issued "Toward a National Science Policy for Canada"
1969	Medical Research Council (MRC) established as a separate entity
1970-77	Reports by Senate Special Committee on Science Policy (Lamontagne)
1971	Creation of Ministry of State for Science and Technology (MOSST)
1978	Creation of Natural Sciences and Engineering Research Council (NSERC)
1978	Creation of Social Sciences and Humanities Research Council (SSHRC)
1984	Wright Report
1985	Government issued Nielsen Task Force report on science and technology
1987	National S&T Policy approved by federal, provincial and territorial governments
1987	Creation of Council of Science and Technology Ministers (CSTM)
1987	Approval of S&T Strategy and Decision Framework by federal government
1987	Creation of National Advisory Board on Science and Technology (NABST)
1988	MOSST merged as a branch into Department of Industry, Science and Technology
1989	"The Halifax Declaration: A Call To Action" issued by National Forum of Science and Technology Advisory Councils
1990-94	NABST issued a series of reports, including:
	Lortie Report (1990) — focussed on management of federal labs
	Nicholson Report (1991) — focussed on need for Canada to change
	Janson Report (1993 and 94) — focussed on S&T priorities
1991-92	Reports issued by government's Prosperity Initiative — addressed S&T challenges
1992	Closure of Science Council of Canada
1993	"MOSST Branch" fully integrated into new Department of Industry
1994	Government announced S&T review in February 1994 Budget speech

Source: Adapted from 1994 Report of the Auditor General of Canada.

cope with the situation. The history of the Ministry of State for Science and Technology (MOSST), founded in 1971, illustrates the rambling and apparently unsuccessful approaches of successive governments. In the eighteen years of its existence, there were no fewer than thirteen ministers at its head and it was reorganized five times. In 1988 it became a mere branch in the Department of Industry, Science and Technology. That department's name was significantly changed in the 1990s into the Department of Industry, which does boast a junior minister, the Secretary of State, Science, Research and Development, who also occupies the post of Secretary of State, Western Economic Diversification.

THE REVIEW

When the Liberals under Jean Chrétien came to power in October 1993, they sought, as noted already, to re-examine the role and performance of the federal government with respect to science and technology. The impetus was not only the irresistible urges afflicting all new brooms, but also worldwide developments: the information revolution, globalization, and increased emphasis on economic competitiveness. It was clear that a strong S&T base was the *sine qua non* of Canadian economic growth and prosperity. Furthermore, the 1994 *Report of the Auditor General of Canada* — Parliament's independent watchdog over the expenditures and performance of various government departments and agencies — included a scathing review of the government's performance in S&T (Auditor General, 1994). It made imperative an examination of existing policies and structures. As a consequence, the Minister of Industry launched the review which is the subject of this paper.

The Industry Canada initiative was followed by the National Advisory Board on Science and Technology (NABST) being asked by the Prime Minister to undertake an independent assessment of both the public consultations and the intragovernmental reviews. NABST worked much more expeditiously than the lumbering Industry Canada review process and its report, *Healthy, Wealthy and Wise: A Framework for an Integrated Federal Science and Technology Strategy*[5] appeared long before the review process was completed. It was therefore available to the drafters of the Industry Canada analyses and recommendations. But its content was only advisory, whereas the range of review reports by Industry Canada have behind them the authority of the Minister and other Cabinet members.

The series of documents comprising the review findings and recommendations fall into three parts: *A Federal Strategy*, which is the centrepiece; *Industry Portfolio's Action Plan* indicating what the eleven agencies under the aegis of Industry Canada, such as the Natural Sciences, and Engineering Research Council (NSERC), the Social Sciences and Humanities Research Council (SSHRC), the National Research Council (NRC), or Statistics Canada, proposed to do to achieve the government's goals; and a series of volumes performing the same task with respect to the various government departments which have more than a casual interest in S&T.[6]

The new S&T strategy seeks to strengthen Canada's performance, while reducing government expenditures in light of its deficit-cutting commitments, through a series of sometimes interrelated measures. The overriding emphasis of virtually all the volumes comprising the reports is the impact of S&T on the economy and specifically on job creation. A major thrust emphasizes the need to narrow the gap between public and private activities; here, the principal mechanism is a series of partnerships between the two sectors. It is also proposed to strengthen S&T by making the internal governmental efforts more accountable, efficient, and better co-ordinated. Many departments of the federal government engage in and/or finance research but there is insufficient co-ordination between them. The volumes listed in Note 1 address this problem and make numerous suggestions for improvement.

Despite the oft-repeated and variously reiterated reference to the need to draw on partnerships with the private sector, the strategy papers do allow that the government ought to provide direction in at least certain sectors of the S&T enterprise. A succinct way of identifying the ideas put forward here is to note the subheadings of the sixth chapter of *Science and Technology for the New Century* (Industry Canada, 1996*b*), which deals with "Providing Direction": increasing the effectiveness of federally supported research, capturing the benefits of partnership, emphasizing preventive approaches and sustainable development, positioning Canada competitively within ... international regulatory standards and intellectual property regimes, building information networks, extending S&T linkages internationally, and promoting a stronger science culture (ibid., 1996*b*, pp. 22-34).

Characteristically for Canada, in the light of the story told by Table 2, considerable importance is also attached to "creating new institutions and mechanisms of governance" (ibid., pp. 15-21). The latest reshuffling of the deck involves the abolition of the National Advisory Board on Science and Technology which, as Table 2 shows, was created only in 1987. It was replaced by a new body, with more limited scope, the Advisory Council on Science and Technology. "... NABST's more public functions," it is argued, "those related to mobilizing the broader scientific community and influencing Canada's attitudes toward S&T, would be best served by institutions at arm's length from government" (ibid., p. 15). To enhance the presence of S&T in the decision-making fora, the Economic Development Policy Committee of Cabinet was charged with the task of formally reviewing federal performance in S&T and making recommendations to Cabinet on the government's S&T priorities (ibid., p. 16).

Underlying the whole exercise was the premise that job creation and economic growth (competitiveness) now and in the next century can only be built on a strong technological base. Noting that the OECD has found Canada suffering from an "innovation gap", the minister of industry and the strategy papers propose to narrow it through "action plans" designed to improve the government's own performance and, particularly, to improve the ability of the private sector to develop and commercialize technology. Although lofty goals are identified ("sustainable job creation and growth, an improved quality of life and the advancement of knowledge") (Industry Canada, 1996*a*, p. 1). an enormous weight is placed on the job-creation potential of S&T, primarily through aiding the private sector to apply available technology more effectively and widely.

Curiously, the principal instrument for this goal is only obliquely noted in the reports but was identified in the budget introduced before the strategy papers were released. It is the "Technology Partnerships Canada" program which provides repayable funds for firms engaged in promising research and development in high technology projects. A $250 million investment fund was created, to support new ventures in three fields: environmental technologies; strategic "enabling" technologies (like biotechnology, some information technologies, advanced manufacturing technologies, etc.) which make industries more productive; and aerospace and defence industries.

Having sketched the main directions of the reports, we can now add some meat to this skeletal account, as we evaluate the government's review.

EVALUATION

The principal benefits of this colossal enterprise are that it clearly identified the goals to be pursued by the government in S&T, that it facilitated the creation of one or two useful policies and programs and, above all, that it acted as a consciousness-raising experience both inside and outside the government. Although one may not agree with the extremely heavy emphasis on applied and competitiveness-oriented S&T in the reports, at least *some* hierarchical ranking of priorities emerged from the process. Duplication of services and programs was also identified, and each department and agency involved in S&T became much better aware of what related activities were afoot elsewhere in the government and of the overall programs touching S&T.

It is still too soon to assess the results of most of the good intentions which luxuriate in the volumes of the report, since it takes a long time for organizational and programmatic blueprints to mature and become realized. But in one area, that which receives the most publicity, some results are becoming evident. Thus, the Technology Partnerships Canada program has enabled Canadian firms to compete effectively in the world economy because of enhanced S&T performance. Among the industries benefiting from the loan fund are aircraft manufacturers (notably Bombardier and Pratt & Whitney), Spar Aerospace for the development of its robot "Canada Hand" — a special-purpose dexterous manipulator to be used in space activities — the pulp and paper industry, and Ballard Power Systems Inc. — a developer of an environmentally friendly fuel cell for powering vehicles, working with major car manufacturers and developing a partnership with the German automobile giant Daimler-Benz (Damsell, 1997).

There is also evidence that closer co-operation among related departments and agencies is coming into being, possibly as the result of the impetus provided by the review process. At any rate, there is some indication that the government is beginning to place more value on research. Extension of permanent funding to the Networks of Centres of Excellence and the creation of the Canadian Health Services Research Foundation and of the Canada Foundation for Innovation are examples. Similarly, other new networks comprising various university research sites, government, not-for-profit organizations and private firms are emerging. One example is the creation of five research and training networks, funded by the Social Sciences and Humanities Research Council (SSHRC), focusing on education, employment, and life-long learning (SSHRC and Industry Canada, 1996).

These fragmentary signs notwithstanding, the situation is threatening, in part because of the pervasive consequences of draconian deficit-reducing measures. These had serious psychological consequences for the review process and they are weakening and distorting ongoing activities in S&T. On the psychological side, they led many of the government departments involved in the review to protect their shrinking resources either by being unwilling to co-operate with other agencies or by simply opting out of, or reducing, their S&T involvement. "Science", in the words of one observer, "is still viewed as a 'cost', not an investment". The climate of deficit reduction also led to the popularity of partnerships, particularly with the private sector. While these can have great merit, they also cause concern: they tend to deflect money and attention from basic research, steer effort towards marketable activities, and diminish the freedom of action of the granting agencies. The president of the Medical Research Council of Canada has noted that "[e]very partnership we enter into, while generating new dollars for research, also requires dollars from our base budget, and that takes money away from our 'core' business, basic research" (Friesen, 1997, p. 1). Only NSERC has made a conscious decision, in 1995, that research partnerships should not direct funding from basic research.

While an evaluation of the review process and its aftermath cannot ignore the prevailing economic circumstances, it must focus on the documents themselves. And here, alas, there is cause for considerable concern, primarily for two sets of reasons: their general tone, which reveals a flawed approach, and the manner in which they deal with specific issues.

GENERAL CONCERNS

Lack of Precision. Statements abound like "[W]e must continuously build on our efforts to use science and technology for the well-being of Canadians" (Industry Canada 1996*b*, p. 6) or "[F]or all concerned, it is time to repurpose (sic) intergovernmental relations in science and technology." No specific targets are identified, no sums specified or no dates of completion indicated. This is troubling in the light of the 1994 Auditor General's report which catalogues innumerable instances of successive governments failing to follow through on promises and pious hopes voiced in previous inquiries into science policy. The Auditor General called for precise benchmarks, yardsticks, and sums to be expended — a plea totally ignored in the 1996 review. This absence contrasts strikingly with the much greater specificity of the recommendations contained in *Realising our Potential: A Strategy for Science, Engineering and Technology*, the British review of the United Kingdom's S&T policies (United Kingdom, 1993).

Disingenuousness. The aura of motherhood infusing the review is matched by a certain disingenuousness, if not downright misrepresentation. Thus, the claim made in the budget brought down more or less at the time the report was published, that $250 million of *new* money would be spent on Technical Partnerships Canada masked the fact that a good proportion of these funds was created by reallocating resources which had previously been earmarked for the Defence Industry Productivity Program whose abolition freed $144 million a year (McKenna, 1996*a*).

At a different level, the review's assertion that it has responded to the advice of the Auditor General's report and to NABST's *Healthy, Wealthy and Wise* is, in significant respects, quite spurious. We have already noted the absence of precisely defined goals, as sought by the Auditor General, and shall show below that the most important recommendations of the two documents just mentioned were in fact rejected.

Shallowness and Superficiality. A failure to probe deeply into questions raised undermines the usefulness of the reports. The boast is made, for instance, that twenty-two researchers who were born or educated in Canada or who worked here, received the Nobel Prize (Industry Canada, 1996*b*, p. 3). But the fundamental question of why so many work abroad and why fewer than a handful still remain in Canada is ignored.

Likewise, the ubiquitous and urgent promotion of partnerships with the private sector fails to address the question of how such hybrid arrangements are likely to affect the all-important pure research. Indeed, unlike *Realising our Potential*, the Canadian review, while paying lip service to the necessity of maintaining disinterested inquiry in science, fails to explore the relationship between pure and applied research, and the implications of this relationship for the direction of discovery and for the training of researchers. Indeed, the implicit placing of pure research on the back burner bodes ill for the future of science in Canada.

Discontinuity. As part of the review process, the Department of Industry organized extensive consultations with hundreds of interested parties throughout Canada. Regional meetings were held, culminating in a final summit in Ottawa. The gatherings were carefully orchestrated, even including an electronically conducted set of votes on a number of greatly oversimplified questions. This gimmickry, of course, led nowhere, but was no doubt intended to convey the impression that genuine consultation was at work. Whatever the merits of some of the meetings, they did produce a large number of ideas and suggestions and many useful insights. The authors of the review documents had a vast amount of raw material to draw on in sifting out their ideas. In addition, as we noted, and as is evident in Table 2, previous studies, inquiries, and reports furnished a mass of documentation illuminating the problems at hand. The very recent Auditor General's and NABST reports were particularly relevant and topical.

It is therefore quite remarkable that the widespread consultations are virtually ignored throughout the reports and are simply banished to a one and a half page ghetto (Industry Canada, 1996*b*, "Annex: Listening to Canadians", pp. 36-37). By contrast, the British report bristles and is interlaced with citations and other references to relevant documents and explicitly weaves diverse suggestions into its text.

While the foregoing features sap the effectiveness of the report, they pale in significance next to other, substantive blemishes.

SPECIFIC ISSUES

Science Culture. Nothing is more critical to the effectiveness of a S&T policy than the underlying values of the host society, particularly with respect to science. This dimension, as we shall see, is particularly germane in Canada. While the report devotes one page (page 7) to the topic, it is filled with generalities and motherhood statements. Essentially, the report argues that education is important and that (again) partnerships between business and scientific institutions are essential. The only concrete factor identified in the *Science and Technology for the New Century* is SchoolNet, the government-supported scheme linking schools, colleges, and libraries to the Internet. No other specific element related to a science culture is identified or proposed.

Government and the Science Community. Numerous references are made in the report to the need for independent advice on issues arising in science policy. But only one such instrument is identified and proposed and it, alas, is merely a watered down version of NABST. This heretofore useful body (its publications alone more than justified its existence) is to be abolished. Its projected successor, the Advisory Council on Science and Technology (ACST) is to differ from it in a number of important ways: its Chair will be the Minister of Industry, rather than a non-governmental person; it will issue no publications and will not have any responsibilities for mobilizing the broader scientific community and influencing attitudes towards S&T. Its members "will represent the spectrum of Canada's economy, and will come from the forefront of Canadian innovators in S&T and other relevant fields." Appointed by, and responsible to, the Prime Minister, it will, unlike NABST, not be at arm's length from government (Industry Canada, 1996*b*, p. 13). Again, the British White Paper provides a sobering contrast: it specifies a whole slew of advisory and executive bodies available to decision makers in science and technology and maps a plausible consulting regime.

Except for the ACST in which, as we saw, the Minister of Industry is to play a key role, the report does not identify other, more independent advisory groups. On the contrary, it suggests that the government intends to play a more central role in the S&T field than ever before. After listing the government's core S&T activities, STNC ventures onto a decidedly imperialist terrain: "... the government will increasingly emphasize a new role: that of information analyst, knowledge disseminator and network builder" (ibid., p. 10). The question is inescapable, of why, in a period of stringent financial retrenchment, these functions need to be assumed by the government when numerous viable institutions, with better credentials, are already in the field performing them.

Implementation Mechanisms. Among the insights of *Healthy, Wealthy and Wise*, none was as valuable as the observation that an effective science policy requires that S&T have a champion in Cabinet. Accordingly, it suggested that a powerful minister be made responsible for S&T. Furthermore, NABST also recommended that a small science secretariat be established to support that minister's work, headed by a senior civil servant bearing the title of chief S&T advisor (NABST, 1995, pp. 10-11). NABST was no doubt motivated by the knowledge that in a society as indifferent and even hostile to science as Canada,[7] a powerful voice for it had to be inserted at the very centre of political power. The authors of *Science and Technology for the New Century* rejected these absolutely critical suggestions, ostensibly on the grounds that "consolidating responsibility in a central structure creates the risk of separating S&T from the core roles, responsibilities and activities of central departments and agencies", and that in a parliamentary system "ministers must retain the capacity to direct their departments' efforts and resources, including ... S&T, ... and be accountable for the results" (Industry Canada, 1996*b*, p. 16). NABST obviously did not think so, nor did the architects of British science policy and the British White Paper. Instead of benefiting from direct access to the Cabinet through a strategically placed minister, the interests of science and technology are, according to the report, to be channelled through the Economic Development Policy Committee of Cabinet — a body whose heavy agenda and eclectic composition are not likely to give much emphasis to S&T.

Absence of Vision. Compared with the British White Paper, which sought to situate British science and technology in a long-term and universal context, the strategy papers are rather more limited. They, in fact, lack broad vision. This manifests itself at several levels.

In the first place, the authors failed to ask themselves some essential questions or approached them in a superficial manner. Noting, for instance, that less than 5 per cent of scientific knowledge and new technologies are developed in Canada, *Science and Technology for the New Century* argues that the world's knowledge needs to be accessed effectively, (ibid., p. 23) but it fails to ask what niche Canada should strive to occupy in this sense in the global context, or on what comparative advantage it can or should draw in its international interactions. Secondly, it supinely accepts Canada's consumer status in S&T without seeking to identify whether this country ought to occupy new ground. Here again the contrast with the British White Paper is painful to a Canadian reader.

Thirdly, the strategy papers have almost nothing to say on the pivotal question of the appropriate training of scientists and on how it is being affected by the extensive downsizing in this sector by both the federal and provincial governments. Finally, as I mentioned previously, instead of specifying new options and proposing a bold revision of the appropriate machinery of government, they downgrade NABST and propose to replace it with a weaker ACST, subservient to the Minister of Industry and his staff.

THE CAUSES OF OUR DISCONTENT

The foregoing requires that we explore two closely related questions. The first concerns the reasons for Canada, a decidedly modern country, having failed to provide an environment that strongly supports the development and growth of S&T. The second deals with the causes for the rather lacklustre quality of the mid-1990s review of science policy. To cover so vast a subject as the societal underpinnings for S&T requires a large canvas. Exigencies of space deprive us of this luxury and we shall therefore have to resort to something of a sketchy caricature, rather than to a fully developed portrait.

A Colonial and Post-Colonial Mind Set. Canada's formative years were spent in the shadow of what many Anglo-Canadians call the "mother country", the United Kingdom. This meant that until about the Second World War many of the dominant ideas, practices and institutions were borrowed from abroad. Schoolbooks, much of cultural life, many industrial and commercial enterprises, the organization of the professions, and the majority of other features of Canadian life were imported from Britain or at least inspired by British prototypes. An overwhelming proportion of Canadians obtaining graduate degrees did so in the United Kingdom.

Under these conditions, a *borrower mentality* predominated. This meant that more often than not, when something new or particularly complex was needed, it was brought to Canada from overseas. This explains why so many members of the academic and scientific communities were British or at least why they came to Canada from Britain. It was usually easier and cheaper *in the short run* to import science and technology from the older and more experienced communities which had created British Canada in the first place. Until the 1950s, the science community and the world of innovation were for the most part thought of primarily as the domain of non-Canadians. To some extent, therefore, a career in S&T was far from the minds of most young Canadians and Canadian enterprises relied heavily on imported S&T. There were exceptions, of course, but generally most Canadians did not see S&T as "their thing". Science culture in Canada was, consequently, very weak.

The war ushered in profound changes by mobilizing Canada's industrial capacity and, in the end, by reducing Britain to a much diminished polity no longer able to play a dominant role in the world. At the same time, Canada's experience in the global conflagration greatly enhanced the country's sense of independence and national identity. One of the results was that an economic, cultural, and educational burst of creativity smashed the vestiges of colonialism and of the dependence on Britain. Science and technology, while affected by these currents, nevertheless grew at a slower pace than many other sectors.

A Branch Plant Economy. One of the reasons was the growing hegemony of the United States of America which supplanted the United Kingdom as the leader of the Western world and as Canada's principal external partner. Canada's economy grew at an impressive rate but also assumed many of the features of a new colonialism, the branch-plant economy. Much of the country's economic growth and strength, particularly in such vital sectors as the automotive industry or oil and gas production, was driven by enterprises owned and operated by gigantic, predominantly American multinational firms. A salient feature of these organizations was that the research and development functions were performed by the mother companies situated in the U.S. This fact contributes not only to the poor R&D record of Canadian industry, but also to the relatively low commitment, on the part of the Canadian public and hence its governments, to science and technology. A comparison with France, Germany, the United Kingdom and the United States is instructive in this context. In a very real sense, Canadians transformed themselves from being borrowers from the British, to becoming borrowers from the Americans.

The Intellectual Development of French Canada. The two dominant linguistic communities of Canada followed substantially different paths of intellectual development. From the time of the victory of the British over the French in North America in the eighteenth century, until the so-called Quiet Revolution of the 1960s and 1970s, French Canada was a largely isolated, self-contained entity whose social and cultural life was dominated by the Catholic Church. Its central goals were the preservation of the French communities, their language and the Catholic faith, and the principal instrument was the maintenance of an extremely high birth rate. The vast majority of *Canadiens* lived in Quebec, which was an overwhelmingly rural society espousing rural values. Educational institutions, also dominated by the Church, stressed the humanities almost at the exclusion of science. Professional training concentrated on the preparation of lawyers and doctors. The latter were primarily trained for careers as clinicians rather than as researchers. Consequently, about a third of Canada's population was, until the 1960s and the 1970s, profoundly uninterested in S&T and deprived of an adequate educational infrastructure fostering the development of a science and technology culture. While this has contributed to the generally inhospitable climate surrounding S&T in Canada, it is important to note that for the last thirty or forty years or so Quebec has fully caught up with the rest of the country and indeed, in some areas — the aerospace sector, for instance — it has become a Canadian and even worldwide leader.

A Resource-Based Economy. For most of the country's history, Canadians have participated in international trade as hewers of wood and drawers of water. Exports of agricultural products and raw materials were, until the Second World War, much more important than manufactured goods. And they are still critical now, although the economic profile is more balanced. While technology now also plays an important role in the resource sectors, it was primarily in industry that S&T flourished. The structure of the Canadian economy has thus also contributed to the limited preoccupation Canadians have had with research and its application.

The Low Valuation Assigned the Intellectual Side of Life. This situation is much more evident in English than in French Canada. Anglo-Canadians generally accord a quite low esteem to the intellectual side of life, whereas Quebecers appear to respect and value it more. The difference in treatment the eggheads receive in the French and English media is telling. Generally, outside Quebec, academics, artists, writers and the like (I include scientists and inventors) are either ignored or viewed with a certain amount of suspicion. There are many reasons for this: Canada's historical evolution, particularly its pioneer past; the ubiquity of the business ethos; widespread materialism; admiration for the decisive *man* of action are some, but there are many more. People who devote their lives to ideas are not seen as making a particularly valued contribution to society. Thus scientists, except perhaps the few who attain extraordinary international recognition (receiving the Nobel Prize, for instance which, after all, is accompanied by a large financial award) are not considered to be role models or suitable subjects for particular public recognition. This being the case, intellectuals, scientists and artists (and their views) tend to be neglected or brushed aside by the major power brokers, including politicians.

A Cottage Culture. Although this point may invite ridicule, it does have some validity despite constituting only a minor factor in our argument: the considerable natural beauty of so much of Canada takes people's minds off the more abstract, ethereal realities of their existence. That so many Canadians, particularly in central Canada, own cottages heightens this tendency. A cottage and its accoutrements soak up a very considerable amount of time and energy, claiming constant ministrations from their handyman owner, so that this hapless individual and his kin have little time left for other pursuits or concerns. The care lavished on getting to the country and back, maintaining one's place, connecting and disconnecting water pipes, mollycoddling the boat(s) may well provide high quality time, but it is time taken from other activities, including serious reading, thinking or keeping abreast of developments in fields such as science.

It must, however, be admitted, that while the foregoing is important, the cottage experience also introduces many Canadians to nature and to natural history and the lore studied by the biological and earth sciences.

Regional and Ethnic Politics. Few countries are as obsessed as Canada with questions of national unity, identity, and of the respective jurisdictions of the central and provincial governments. More recently, these preoccupations were also joined by concerns over the entitlements of the aboriginal population, women, the disabled, and groups espousing certain lifestyles. Together all these themes have absorbed and continue to absorb so much public care and to claim so much of the attention of activists, the media and policy makers that relatively little is left over for other matters. The cultural domain, education, and a host of other important areas of human experience, *including science* tend, therefore, to be short changed.

Small Country Status. Canada's status as a small country, given to peacekeeping rather than to sabre rattling, contributes to the neglect of science and technology. Countries like the United States, France and the United Kingdom, with rich histories of foreign military involvements and worldwide strategic interests, have allocated a considerably larger proportion of their human and monetary resources to the so-called "defence industries". These, as is widely appreciated, have played a major catalytic role in scientific discovery and in the invention and perfecting of new technologies. Canada has, for the most part, avoided becoming heavily involved in launching major, defence-driven scientific work. In the few instances when it did flirt with them, it did not stay the course. The development and then abandonment in 1959, of the state of the art supersonic, all-weather interceptor jet aircraft, the Avro Arrow, is a dramatic illustration. The decision to scrap the project led to the firing of 14,000 skilled workers and to the loss to Canada of a very large number of scientists and engineers who went to the United States.

Bureaucratic Culture. Canada's stance vis-à-vis S&T, and particularly the lamentable quality of the policy review, are also influenced by the nature of the bureaucratic process and by the folklore of the federal public service. This dimension, which is of critical importance not only to science policy but also to many other aspects of government in Canada, is normally ignored; I will, therefore, devote somewhat more space to it than to many of the other items discussed here.

Canada has created and benefited from a superb public service but there are numerous indications that its quality has been deteriorating. There are still excellent officials in place, of course, performing admirably, but the system as a whole exhibits dangerous weaknesses.

The S&T review process serves as an excellent illustration. Although, as already noted, a certain show was made of basing it on extensive consultation, it was in effect the officials in the Department of Industry (and some consultants hand-picked by them) who crafted the reports and who shaped the recommendations. These, as we saw, failed to reflect the most important suggestions offered by NABST and the Auditor General, and adequately to integrate most of the useful ideas gathered in the regional and Ottawa consultative meetings. The reason lies in the mind-set prevailing in the bureaucracy and in the tendency, among many of its leading members, to reject any changes likely to weaken their own power and influence.

This stance is not merely self-serving. It also results from a widespread conviction of the superiority brought to policy issues by the federal bureaucracy. The latter, it is assumed, is on the whole more disinterested and more far-sighted than private interests, lower levels of government and a great many politicians. Furthermore, it is also taken for granted or at least tacitly assumed that the way things look from Ottawa is somehow more embracing and superior to the optics emanating from other jurisdictions. The capital is seen as the quintessence of the country, and its priorities and agendas are deemed to represent the highest aspirations of Canada as a whole. Ottawa's definition of the public interest is therefore taken to be more altruistic and superior to that espoused by others.

In arrogating to themselves immense powers and vast spheres of influence, the mandarins believe themselves to be serving the public interest better than can be done by anyone else. They thus measure up, in their own eyes, to the most exalted criteria of an enlightened and competent public service. The country deserves the best, and it is the superior competence of the federal bureaucracy which ensures that it shall get it.

While there are without doubt many exceptions to this characterization of the senior levels of the public service, it has been amply evident in various policy areas including, as noted above, in that of S&T. The strategy papers, as I have shown, failed to recommend the creation of a science secretariat with a voice in Cabinet independent of the Minister of Industry, they replaced NABST with a less powerful group chaired by that Minister, they did not provide a blueprint for an effective consultative network of consultations drawing on the rich non-governmental S&T organizations available in the country. The recommendations considerably enlarge the oversight of science by the Minister and his staff. They serve their boss — the Minister — and themselves — his advisers — more than they do science, technology and the intellectual community as a whole. I can only conclude that the authors were engaged in the time honoured, but usually myopic, exercise of turf protection and turf extension.

Political Naiveté of the Science Community. One more reason needs to be invoked explaining why the societal context has inhibited the creation of a vigorous science culture in Canada. And happily, after what has been a rather sad tale, this factor is amenable to reasonably speedy correction.

Although the science community in Canada has performed quite remarkably well on the S&T front, despite some of the impediments noted earlier, it has been much less effective at the political level. Compared with the public service, for instance, it has not feathered its nest or attracted much by way of perks, individually or for its collective tasks. It is, in fact, not a great exaggeration to assert that it, and particularly the S part, has on the whole been naive, unaware, unorganized, and inept in putting its needs on the public agenda.

There are exceptions, to be sure, and some beachheads exist, but compared to many other occupational and professional interests, the savants have underperformed. Ironically and atypically, at the time of writing, the heads of all three granting councils and of the NRC are highly esteemed members of the science fraternity, a factor justifying mild optimism about the future.[8] But generally, the policy process, even with respect to S&T, has not benefited from the vigorous participation of the practising cadre of Canadian scientists.

Among the many reasons for this, one stands out: scientists often do not trouble to take the time to understand the political process and/or to influence it, for the case for science appears to its practitioners to be so self-evident that it is unnecessary (and distracting) to make it; the scientific mind is often repelled by the messy way in which political and administrative decisions are made; there is only an infinitesimal number of scientists in the policy community; officials in positions of authority tend generally to pay only casual attention to what scientists think; procedures for channelling S&T advice to the government are primitive, underdeveloped and underused; the science community has been fragmented, composed as it is of components from university, government and private firms; the need to become involved, lobby, and reach out to the political process has been largely ignored; machinery for injecting the voices of S&T into the decision-making process has been rudimentary and imperfect.

Given this state of affairs, it is not wholly misleading to conclude that in contemplating their relations with Caesar, the science savants are justified in repeating Pogo's most famous and most oft-cited dictum: "We have met the enemy and it is us."

A WEE PHOTON OF LIGHT AT THE END OF THE TUNNEL

Fortunately, however, remedial action is in sight, at least in one domain. Two organizations have entered the field seeking to improve matters. One is the Partnership Group on Science and Engineering (PAGSE, pronounced, intriguingly, "pages"), the other is the Canadian Consortium for Research (CCR).[9] Both include scientific organizations, grouping professionals in various disciplines as well as larger, more general organizations like the Royal Society of Canada. PAGSE, confined to scientific organizations, seeks to inform, educate, and awaken the political class, primarily parliamentarians, the Cabinet, public servants, and the media. CCR, which includes student organizations as well as science bodies, focuses more on lobbying as such. Results, so far, appear to be modest but at least mechanisms are in place and breathing, capable of becoming an effective interlocutor for S&T. While this is not much, it is a beginning.

Another faint hope is nourished by the government's (and the Minister of Industry's) intense desire to foster partnerships between businesses and scientific organizations in the cause of applying science and technology to the shoring up of the Canadian economy. The Canada Foundation for Innovation, announced in the 1997 budget, is a case in point (Canada, 1997). If all goes well, the foundation could generate up to $2 billion from public and private sources in investments in the research infrastructure (ibid., p. 4). While the government's obsession is with applied rather than pure research, scientists have been known to direct their work towards gratifying, at least in part, their thirst for curiosity-driven research. Between the better organization of the S&T community, the government's interest in enhancing job creation, the success of Ottawa in deficit reduction, and the late 1990's improvement of the Canadian economy, there is some hope that the decline in support for science may be halted and reversed. But as the whole tenor of this piece indicates, these hopeful signs will, even if followed by improvements, come to naught until some of the underlying and long-term factors impinging on the S&T regime noted already are altered. And that will take a long time, and a concerted effort by many hands, both inside and outside the world of science and technology.

NOTES

Among the many kind members of the science community who have given me unstinted advice and encouragement, three deserve special mention: Dr. Agnes M. Herzberg, Queen's University, originally goaded me into delving into the thicket of science policy in Canada and then on numerous occasions relentlessly pressed me into putting my resulting reflections into written form. Dr. Paul Hough, Executive Director of the Canadian Federation of Biological Sciences, creatively led me to many elusive sources and saved me from committing some gaffes in the reading of events. Neither, needless to say, necessarily shares my views or bears any responsibility for the shortcomings of this paper. The same dispensation applies to Michael Dence, then the Secretary of the Science Academy of the Royal Society of Canada who, when I embarked on this project, also generously provided me with essential documents for my review of the review.

1. Industry Canada, *Science and Technology for the New Century* (henceforth abbreviated *STNC*); *STNC: Industry Portfolio's Action Plan*; *Highlights of Departmental S&T Action Plans in Response to STNC*. These are the principal reports, but in addition Industry Canada released, at the same time, a series of more narrowly focused reports under the following titles: *Highlights of the Departmental S&T Action Plans in Response to STNC*; *STNC: Agriculture and Agri-Food Canada's Action Plan*; *STNC: Environment Canada's Science and Technology: Leading to Solutions*; *Fisheries and Oceans Action Plan: STNC*; *S&T at Work for the Health of Canadians: The Science and Technology Action Plan of the Federal Health Portfolio*; *STNC: A Framework for the Human Resources Management of the Federal Science and Technology Community: STNC*; *Human Resources Development Canada: Using Science and Technology to Better Serve Canadians: Human Resources Plans in Response to STNC*; *Natural Resources Canada: New Directions in S&T*; *Transport Canada Action Plan: S&T for the New Century*; *Defence Research and Development: STNC*. These publications comprise the government's new S&T strategy and are sometimes referred to in what follows as the "strategy papers" or "the reports".
2. Communication from Paul Hough, 26 May 1997.
3. Strauss (1996, 1997) and Spears (1997*a*, *b*). It is revealing that the continuation overleaf of this front-page story is entitled "Science: Making Money Most Important".
4. For a brief list and summary of the reports of *some* of the relevant review bodies, see "A Selected Bibliography of Major Federal S&T Policy Reports" in Industry Canada (1994).
5. NABST, Ottawa, April 1995. For a succinct description of NABST's activities and a list of its other reports, see Industry Canada (1995, vol. 2, pp. 67-69).
6. As indicated in Note 1, the principal volume is identified below as *STNC* and the ensemble of the reports as the strategy papers or the reports.
7. This aspect of Canada is developed in the next section.
8. Dr. Thomas Brzustowski heads the National Science and Engineering Council, Dr. Arthur Carty is in charge of the National Research Council,

Dr. Henry Friesen runs the Medical Research Council, and Dr.Marc Renaud took over the Social Science and Humanities Research Council in 1997. They all have impeccable academic backgrounds and are at home in the corridors of power. C.P. Snow would have seen them as denizens of both of the two cultures.

9. The CCR is the new name for the invigorated National Consortium of Scientific and Educational Societies (NCSES), founded in 1976.

REFERENCES

Anderson, E. (1997*a*), "Research Cutbacks Drive Out Top Geologist", *The Globe and Mail*, 20 October.

________ (1997*b*), "Canadian Science Going the Way of Dinosaurs", *The Globe and Mail*, 27 October.

Auditor General of Canada (1994), *Report of the Auditor General of Canada*, Ottawa: Supply and Services Canada.

Canada. Her Majesty the Queen in Right of Canada (1997), *Canada Foundation for Innovation*, Ottawa: Department of Finance.

Damsell, K. (1997), "Ballard Shares Electrify Market", *The Financial Post*, 11 October.

Friesen, H. (1997), "A Report Card on the Well-Being of Health Care Research in Canada: Choices to be Made", address to "The Well-Being of Canada Symposium," of the Royal Society of Canada, 22 November.

Industry Canada (1994), *Resource Book for Science and Technology Consultations*, 2 vols., Ottawa: Supply and Services Canada.

________ (1996*a), News Release*, 11 March.

________ (1996*b*), *Science and Technology for the New Century: A Federal Strategy*, Ottawa: Supply and Services Canada.

Liberal Party of Canada (1993), *Creating Opportunity: The Liberal Plan for Canada*, Ottawa: The Liberal Party of Canada.

May, R.M. (1997), "The Scientific Wealth of Nations", *Science* 275:793-96.

McIntosh, A. (1997), "NRC Hopes Bonuses Will Slow Brain Drain", *Ottawa Citizen*, 16 October.

McKenna, B. (1996*a*), "Superfund to Boost Science, Technology", *The Globe and Mail*, 5 March.

________ (1996*b*), "Feds to Revamp Science Plan", *The Globe and Mail*, 11 March.

NABST (1995), *Healthy, Wealthy and Wise: A Framework for an Integrated Federal Science and Technology Strategy,* Ottawa.

OECD (1997), *Main Science and Technology Indicators* 97-1. Paris.

SSHRC and Industry Canada (1996), "Education and Training Research Networks Established by Federal Research Council", *News Release*, 5 December.

Spears, T. (1997*a*), "Canada Lets $70-Million Research Facility Die", *Ottawa Citizen*, 1 February.

________ (1997*b*), "Brain Drain: Part 2", *Ottawa Citizen*, 10 May.

Strauss, S. (1996), "Scientists Try to Save Atomic Facility", *The Globe and Mail*, 1 September.

________ (1997), "AECL News Good, Bad", *The Globe and Mail*, 17 January.

United Kingdom (1993), *Realising our Potential: A Strategy for Science, Engineering and Technology*, Cm 2250, presented to Parliament by the Chancellor of the Duchy of Lancaster.

PART X

DISCUSSION

CHAPTER 25

The Federal Strategy for Science and Technology in Canada and Statistical Measurement

F.D. Gault

INTRODUCTION

Statistics Canada is building a program of measurement of science and technology (S&T) activities throughout the economy. This program, funded by Industry Canada for four years, is one of the substantive outcomes of the federal science and technology review and it is a step towards monitoring the progress of the federal strategy announced in March 1996. This report emphasizes how the products of the program can be used for this purpose, and to tell Canadians how the public purse is spent.

THE FEDERAL SCIENCE AND TECHNOLOGY STRATEGY

In 1993-94, the federal government committed itself to managing its science and technology portfolio more strategically and in the 1994 Budget (Martin, 1994), it announced its intention to "put in place a true strategy, one with real priorities, real direction, and a real review of results." As part of this, the *Science and Technology Review* was launched in June 1994 and included a review by federal departments and agencies of their activities, an independent review by the National Advisory Board on Science and Technology (NABST) and public consultations across the country.

The federal S&T review reported on 11 March 1996 and that report (Government of Canada, 1996) put forward three goals for federal S&T activities: sustainable job creation and economic growth; improved quality of life; and advancement of knowledge.

With these as goals, three core activities were identified for the federal government: (i) funding and performing scientific research to support the mandates of departments and agencies; (ii) supporting research in universities, colleges, hospitals and other non-governmental research institutions and Networks of Centres of Excellence; and (iii) supporting private sector research and development.

Finally, a common framework of operating principles was presented. The seven principles were: (i) increasing the effectiveness of federally supported research; (ii) capturing the benefits of partnerships; (iii) emphasizing preventative approaches and sustainable development; (iv) positioning Canada competitively within emerging international regulatory standards and intellectual property regimes; (v) building information networks (the infrastructure for the knowledge economy); (vi) extending S&T linkages internationally; and (vii) promoting a stronger science culture.

While the federal S&T strategy went some way towards meeting the objectives set in the 1994 Budget, it did not provide government-wide priorities. The strategy did acknowledge that "a more rigorous collective view of priorities" was required and went on to say that a mechanism had been established which would eventually "make recommendations to Cabinet on the government's S&T priorities."

One of the outcomes of the S&T review was the funding by Industry Canada of a statistical program, for four years, to develop useful indicators of activity and to move towards a framework to tie them together into a coherent picture of science and technology in Canada (Statistics Canada, 1994). The picture created by these indicators would provide a background for federal government departments and agencies against which to measure their effective application of the seven operating principles as they engaged in the three core activities as part of achieving the three goals of the strategy. More broadly, the picture would be a move towards showing how the Canadian S&T system worked.

THE ROLE OF STATISTICS CANADA

There are four questions that require answers if the progress of the science and technology strategy is to be monitored. (i) What does the federal government spend on S&T? (ii) Where does it spend its S&T resources? (iii)What does it spend its S&T resources on? (iv) What does it get in return for its S&T spending?

Some of these questions can be answered, some require new measures and some require measures that have yet to be developed. Even the first question is not as straight forward as it may seem.

What Does the Federal Government Spend on S&T?

Figures for federal spending on science and technology for the last ten years are given in Table 1, as well as for spending on R&D.[1] The figures show the magnitude of spending and the spending as a percentage of the federal budget.

These figures, combined with foregone tax revenue of about a billion dollars, prompted questions about what the government received for this money, which led to the federal review of science and technology in 1994.

Where do the figures come from? To produce these figures, a questionnaire is sent to every department or agency of the federal government. The most recent one was mailed in November 1996 and it seeks information for 1995-96 actual expenditure, 1996-97 forecast expenditure and 1997-98 estimated expenditure. When the questionnaire is received, the respondent has to take a number of decisions. The first decision is to identify the percentage of the budget, for each of the three years, which is allocated to S&T, and the same is done for the allocation of full-time equivalents (FTEs), to get the human resource allocation. For expenditure, these are the figures that give the results in Table 1. However, their identification is not simple as it does require an informed review of every program, unless the entire budget can be classified to science and technology.

Once the S&T expenditure is identified, there is a second decision required to divide it into R&D and related scientific activities (RSA). Finally, all of the expenditures are assigned to the natural sciences and engineering (NSE) or to the social sciences and humanities (SSH).

Table 1: Federal Expenditures on R&D and S&T,[1] 1987 to 1996

Year	*Budgetary Main Estimates*[2]	*S&T*		*R&D*	
	(in billions of dollars or percentage of budget)				
	$	$	%	$	%
1987	126	4.5	3.6	2.7	2.1
1988	133	4.8	3.6	2.9	2.2
1989	143	5.1	3.5	3.1	2.2
1990	148	5.5	3.7	3.3	2.2
1991	158	5.8	3.7	3.5	2.2
1992	161	5.8	3.6	3.6	2.2
1993	161	5.9	3.7	3.6	2.3
1994	161	5.7	3.6	3.6	2.3
1995[p]	164	5.6	3.4	3.4	2.1
1996[e]	157	5.4	3.4	3.2	2.0

Notes:

1. *Federal Science Activities*, Catalogue No. 88-204, Statistics Canada.
2. Part 1, *Government Expenditure Plan*, Values.

p: preliminary figures
e: estimated figures

Quality and timeliness of the response. The figures reported in Table 1 and in *Federal Science Activities* (Statistics Canada 1995, 1996*a,b,c,d,e*) require the commitment of resources on the part of the responding departments and agencies and on the part of Statistics Canada, which maintains the definitions and controls the data quality. These figures identify the magnitude of the direct spending of the federal government on S&T.

In addition to direct spending on S&T, there is the foregone tax revenue from the Scientific Research and Experimental Development (SR&ED) program which is recorded by Revenue Canada. In the discussion of the following three questions, only direct spending is considered.

Where Does it Spend its S&T Resources?

Federal S&T activity is performed within the department or agency, or it is performed externally, in response to grants and contracts from the government. In both cases, a geographical location is assigned and comparisons can be made of the federal S&T spending in each of the provinces and the territories (Statistics Canada, 1996*a*).

What Does it Spend its S&T Resources On?

To identify where the money goes, the federal survey of S&T activities has been expanded and respondents are asked to assign their S&T spending to one or more of the categories (Office of Official Publications, 1994) in Table 2. The resulting tabulations will answer questions about the total government S&T spending on telecommunication, for example. As departments and agencies become more practised in responding to these questions, the answers will provide a better picture of what the government spends its S&T money on. This is a first step towards understanding spending priorities and is an input for discussions on priority setting and accountability.

Table 2: Socio-Economic Objectives

1. Exploration and Exploitation of the Earth
2. Infrastructure and General Planning of Land Use
    - 2.1 Transport
    - 2.2 Telecommunication
    - 2.3 Other
3. Pollution and Protection of the Environment
4. Public Health
5. Production, Distribution and Rational Utilization of Energy
6. Agricultural Production and Technology
    - 6.1 Agriculture
    - 6.2 Fishing
    - 6.3 Forestry
7. Industrial Production and Technology
8. Social Structures and Relationships
9. Exploration and Exploitation of Space
10. Non-Oriented Research
11. Other Civil Research
12. Defence

What Does it Get in Return for its S&T Spending?

The survey of federal S&T activities will also give some of the answers to this question, other answers are found in other Statistics Canada surveys, and some ways of answering this question are still under consideration.

All of the answers have to be interpreted in the context of the federal strategy, the goals, the core activities and the operating principles which it embodies.

Answers from the Federal Survey. The federal survey, with the addition of the list of socio-economic objectives, allows every S&T activity to be identified with one of the core activities of government stated in the S&T strategy. To answer questions about the seven operating principles, such as capturing the benefits of partnerships, requires further additions which could be made to improve the monitoring of the strategy.

A Broader Picture. So far, science and technology, in this paper, has been limited in meaning to the combination of R&D and related scientific activities. However, science and technology goes well beyond that to include invention, innovation and the diffusion of new technologies, as well as measures of the human resources required for all of these undertakings.

As part of the Statistics Canada initiative, funded by Industry Canada, surveys are being undertaken to measure innovative activities in firms in service industries, work that builds upon earlier measures of innovation by Statistics Canada (Baldwin, Chandler, Le and Papailliadis, 1994), particularly in manufacturing (Baldwin and DuPont, 1996). As part of these surveys, respondents will be asked about government as an influence on innovation in the firm. They will also be asked about their intention to increase or decrease their workforce and the skill levels required if there is a change. These surveys will support the analysis of federal government involvement in innovation through various programs and will provide another dimension to the answer to the question about what the government gets in return for its S&T expenditure.

Other examples of new survey results are the reporting of the Canadian industrial R&D devoted to key technologies, such as software (Statistics Canada, 1996*b*), biotechnology (Statistics Canada, 1996*c*), and pollution abatement and control (Statistics Canada, 1996*d*), and the conducting of a survey of use and planned use of biotechnologies in industry. As well, there are plans to analyze trade in intellectual property and in services related to science and technology, to use the Statistics Canada Input-Output model to examine merchandise trade, to use other Statistics Canada databases to look at job and firm creation, to survey the use of venture capital in Canada, and to survey the use of alliances in all aspects of science and technology activity.

Finally, there is a framework being developed to bring these, and other new indicators, together, to make them easier to understand, as part of the picture of science and technology activity in Canada.

CONCLUSION

New and improved surveys, new analysis and new results, developed over the next three years will provide answers to questions about government S&T spending. Some answers will be straightforward, some will require work which is well understood, and some will require research into the development of new indicators. Taken together, the results of this work will contribute to the monitoring of the progress of the S&T strategy.

NOTE

1. Research and experimental development (R&D) comprise creative work undertaken on a systematic basis in order to increase the stock of knowledge, including knowledge of man, culture, and society, and the use of this stock of knowledge to devise new applications. In addition, there are a number of activities closely connected to R&D which are termed related scientific activities (RSA). Examples of RSA are general purpose data collection, museum services, and education support.

   For the purpose of federal reporting, resource allocation for science and technology is the sum of the allocation for R&D and for RSA.

REFERENCES

Baldwin, J. and Da Pont, M. (1996), *Survey of Innovation and Technology 1993: Innovation in Canadian Manufacturing Enterprises*, Catalogue No. 88-513-XPB, Ottawa: Statistics Canada.

Baldwin, J., Chandler, W., Le, C. and Papailliadis, T. (1994), *Strategies for Success: A Profile of Small and Medium-Sized Enterprises (GSMEs) in Canada*, Catalogue No. 61-523E, Statistics Canada.

Government of Canada (1996), *Science and Technology for the New Century: A Federal Strategy*, Ottawa.

Martin, P. (1994), *The Budget Speech* 22 February, Ottawa: Department of Finance.

Office of Official Publications of the European Community (1994), *The Nomenclature for the Analysis and Comparison of Scientific Programmes and Budgets*, Luxembourg.

Statistics Canada Working Group on the Development of Science and Technology Statistics (1994), *An Information System for Science and Technology, Federal Science and Technology Review, Final Report*, Ottawa: Statistics Canada.

________ (1995), *Federal Scientific Activities 1995-96*, Catalogue No. 88-204, Ottawa: Statistics Canada.

________ (1996*a*), *Federal Government Expenditures on Scientific Activities, 1996-97, Science Statistics*, Catalogue No. 88-001, 20, 4, Ottawa: Statistics Canada.

________ (1996*b*), *Distribution of Federal Expenditures on Science and Technology, by Province and Territories, 1994-95*, Catalogue No. 88-001, 20, 8, Ottawa: Statistics Canada.

________ (1996*c*), *Software Research and Development(R&D) in Canadian Industry, 1993*, Catalogue No. 88-001, 20, 2, Statistics Canada.

________ (1996*d*), *Research and Development (R&D) for Pollution Abatement and Control in Canadian Industry, 1990, 1991 and 1993*, Catalogue No. 88-001, 20, 11, Ottawa: Statistics Canada.

________ (1996*e*), *Biotechnology Research and Development (R&D) in Canadian Industry 1989 and 1993*, Catalogue No. 88-001, 20, 12, Ottawa: Statistics Canada.

CHAPTER 26

Public Policy and Basic Research

R.E. Taylor

INTRODUCTION

Relatively simple statistical problems of physics can still present challenges; sometimes a complex statistical analysis is crucial to the conclusions. As the complexity of the analysis grows, three things may occur: (i) a mismatch between the real problem and the theoretical assumptions under which the methods of analysis have been developed; (ii) there are errors that are not random. These "systematic errors" may be dominant, and the estimation of these is an art; (iii) perfectly adequate statistical analyses are applied to the data, but the question is wrong, and/or ignores pertinent variables. In all these areas, the human element spoils the estimate of the "significance" of the data.

Judging the confidence one should have in a conclusion is more complex than just "looking at the numbers" from a statistical analysis. Models and theories in experimental science often connect with data in a way which helps to sort things out.

FUNDING FOR BASIC RESEARCH

Science has made an unbelievable amount of progress in the last fifty years. Increases in available funding have been driven by World War II, H-bomb politics, Sputnik, missile gaps, and now increased international economic competition. Scientists doing basic research exploit both political and economic problems to justify the most abstract research arguing that knowledge is power, particularly given the progress in technology in the last fifty years.

As public policy shifts towards stressing the economic benefits of basic research, the main difficulty is the time delay between research and application. Unfortunately, politicians and voters tend to take the short view and, therefore, have trouble committing the high levels of support needed for basic research, the kind of research that may change things in a few decades, if at all, but which sometimes change things very much indeed. Governments today are looking for ways to shift support from basic research to those areas of research that look most "relevant" to early technological progress.

GAINS FROM BASIC RESEARCH

The politicians and the general public do not seem to realize the potential gains that can be made from supporting basic research. If one buys an ordinary lottery ticket, the odds of winning are poor, less money being paid out than is paid in. But suppose that lotteries were subsidized and distributed much more money than the total amount of the bets, then buying tickets would be a good investment. This is the sort of lottery scientists believe they are operating, a lottery where the payout is very much greater than the ticket price.

Nowadays, because of the expensive equipment needed for some research, tickets are becoming even more expensive, to the point where some ticket costs for some research must be shared by several countries.

The funding crisis is not confined to expensive science. Growth in funding has definitely slowed in almost all the advanced countries. A lot can still be done within the present levels of support, but we need to adjust our expectations. We have to be more realistic and admit that growth in overall funding is likely to be very lean or even negative for some years to come.

Japan plans to double its spending on basic research over the next five years. If Japanese science makes a breakthrough with important economic consequences, then perhaps other governments can be persuaded that basic science is indeed an engine that drives technology and industry; more support may follow. If this does not happen, then the scientists will have to turn to the statisticians to demonstrate that basic research is worth supporting at the kind of levels we would like to see. This would take us back to statistics in its original form, the collection and analysis of information.

Ralph Gomery (1995, p.14) makes the point that one is used to things that are too small to see with the naked eye and one needs a microscope to look at them; but some things are too big to see. For the latter, one needs a "macroscope". Statistics is an attempt at macroscopy. What one needs most is a focus knob on the macroscope. "If you can build a better macroscope, people will beat a path to your door."

CONCLUSION

Scientists are curious and governments want solutions to problems, problems that scientists believe their research will help in the future. Finding the right level of support is not an easy task. It is to be hoped that in the future, statistics and statisticians can help produce a sensible public policy for science support.

NOTE

This paper is a summary of the speech given at the conference.

REFERENCE

Gomery, R. (1995), "An Unpredictability Principle for Basic Research", the 1995 William D. Carey Lecture, in *AAAS Science and Technology Policy Yearbook 1995,* eds. A.H. Teich, S.D. Nelson and C. McEnaney, Washington, D.C.: American Association for the Advancement of Science, pp. 5-17.

APPENDIX

Saturday, April 20, 1996

Peter Calamai

Bridging moat 'twixt science and citizenry

Outside was a moat and walls built in 1440, when England was perpetually at war and the first book not yet printed. Inside were 40 scientists, worried that knowledge gained over past centuries was threatened by an intellectual Dark Age.

Herstmonceux Castle, donated by an alumnus, is an international study centre for Queen's University in Kingston. It was an indomitable Queen's University statistics professor, Agnes Herzberg, who arranged the three-day gathering in southeast England to focus on communication and miscommunication among scientists, the public, the media and public officials, elected and otherwise.

The depth of that miscommunication was borne home whenever the participants sat down to the castle's beef-free meals; ignorance and misuse of science by politicians has effectively destroyed Britain's beef industry. But why, and could a repetition be avoided?

As always, there is the shadow of C.P. Snow's Two Cultures, a phrase that entered the language in 1959 to describe the gaps between arts and science and religion. If anything, the gaps have widened since, as the handling of risk from mad cow disease illustrates.

Politicians and public servants, overwhelmingly educated in the arts and humanities, crave certainty. For the politician particularly, uncertainty is weakness. Scientists, by contrast, know that certainty is impossible; they are mostly interested in probabilities.

But the scientists are poor at communicating to the public why probabilities are the best we can hope for. So the public seizes on the false certainties promised by politicians and by others with agendas, like the tobacco industry. And we are fed weasel phrases, such as "no scientific proof that" or "no objective evidence of."

The problem, however, goes deeper. Most non-scientists simply don't understand what scientists do and most scientists are terrible at explaining why what they do matters. Of course, sometimes they don't know why, other than to answer unanswered questions.

Questions such as the nature of nuclei in atoms, which turned out to be miniature magnets. When this discovery was made, in the 1940s, there was no practical application for this knowledge. It was the basic or "pure" research now derided by many. Yet the fruits of that labor today sit in every hospital as magnetic resonance imaging systems and helped define the mineral wealth at Voisey Bay through nuclear magnetic resonance.

Since politicians don't understand science or scientists — and therefore can't make rational decisions about spending tax dollars — they seek out surrogate decision-makers, like letting industry pick what research should be funded in universities or government labs. But while there's a powerful business constituency for research into treatments and cures, there's not one for prevention of illness in the first place. Governments also have a weakness for the team approach, overlooking evidence that many of the best scientific researchers work better alone.

From Australia, Britain, United States and Canada, scientists at the Herstmonceux Castle meeting all reported opposition to the very concept of scientific research in their countries. (The mere word "research" in a name can mark an agency for punitive downsizing, so one federal department in Canada is considering changing a "research and analysis" section to "reports and articles" as a more defensible activity.)

For all these concerns, many of the scientists at Herstmonceux had only a superficial understanding of the modern political process. They yearned for a time when an all-powerful science czar needed only to whisper to a key cabinet minister and funding appeared. They failed to research the workings of legislative committees. They seemed not to absorb when one participant accurately described government emphasis today as being on process, not on decisions.

But the scientists are getting angry. In Australia, they turned up with placards and white lab coats in the national capital to protest after the science minister called them whiners. In Canada, the recent non-policy on science and technology from the federal government is actually spurring disparate scientific societies into a united effort.

The moat may still be there but the drawbridge is down and the castle defenders are ready to come out and do battle with words, and votes.

Peter Calamai is *Citizen* editorial pages editor. You can read this column, plus previous Calamai columns, at the Ottawa Citizen Online http://www.ottawacitizen.com

CONFERENCE ON STATISTICS, SCIENCE AND PUBLIC POLICY

10-13 APRIL 1996

Queen's University International Study Centre   Herstmonceux Castle   Hailsham, U.K.

PROGRAMME

WEDNESDAY 10 APRIL

Dinner
18.00 - 19.00

Opening Session
19.30 - 21.00

19.30 Introduction
19.45 Sir John Kingman *University of Bristol*
20.30 Discussion

THURSDAY 11 APRIL

A. Coverage by the Media
8.30 - 10.30

8.30 Introduction
8.35 P.Calamai *The Ottawa Citizen*
9.05 J.C.Bailar *University of Chicago*
9.25 C.C.Heyde *Australian National University and Columbia University*
9.45 Discussion

B. Improvement of Publicity for Science
11.00 - 12.00

11.00 Introduction
11.05 L.L.Conquest *University of Washington*
11.25 Discussion

C. Public Policy I
13.15 - 15.55

a) ***Role of Government Statistical Offices***
13.15 Introduction
13.20 B.A.Bailar *University of Chicago*
13.40 J.F.Gentleman *Statistics Canada*
14.00 Discussion

b) ***Role of Funding Agencies***
14.15 Introduction
14.20 T.A.Brzustowski *Natural Sciences and Engineering Research Council of Canada*
14.40 S.Fortier *Queen's University*
15.00 Discussion

c) ***Role of National Academies***
15.15 Introduction
15.20 C.C.Heyde *Australian National University and Columbia University*
15.40 Discussion

D. Public Policy II
16.25 - 18.00

d) ***Implementation by Government***
16.25 Introduction
16.30 E.A.Gehan *Georgetown University Medical Center*
16.50 I.Krupka *Pest Management Regulatory Agency, Canada*
17.10 J.R.Hemsley *Brameur / Vasari Enterprises Ltd.*
17.30 Discussion

Dinner / Concert
19.00 - 21.00

FRIDAY 12 APRIL

E. Hazards and Risks

8.30 - 10.15

8.30 Introduction
8.35 W.Leiss *Queen's University*
8.55 D.V.Bates *University of British Columbia*
9.15 J.F.Lawless *University of Waterloo*
9.35 Discussion

F. Statistics

10.45 - 12.30

10.45 Introduction
10.50 D.J.Hand *The Open University*
11.10 D.McDonald *Monash University*
11.30 H.P.Wynn *University of Warwick*
11.50 Discussion

G. Scientists and Statisticians

14.00 - 15.45

14.00 Introduction
14.05 D.R.Cox *Nuffield College, Oxford*
14.25 A.T.Stewart *Queen's University*
14.55 Discussion

H. Concerns of Scientists in Public Policy

16.15 - 17.30

16.15 Introduction
16.20 D.J.Bartholomew *London School of Economics and Political Science*
16.40 G.H.Reynolds *Centers for Disease Control, Atlanta*
17.00 Discussion

Banquet

19.00 - 21.00

20.00 P.Milliken *Member of Parliament, Canada*

SATURDAY 13 APRIL

I. Science and the Public Purse

8.30 - 10.30

8.30 Introduction
8.35 J.Meisel *Queen's University*
9.20 Discussion

J. Discussion

11.00 - 12.30

11.00 Introduction
11.05 F.D.Gault *Statistics Canada*
11.25 R.E.Taylor *Stanford University*
11.45 Discussion

We gratefully acknowledge the following benefactors whose generosity has made this Conference possible:

The Davies Charitable Foundation

Falconbridge Limited

U.K. Foreign and Commonwealth Office

Vasari Enterprises Limited

Some anonymous donors are also gratefully thanked.

CONFERENCE ON STATISTICS, SCIENCE AND PUBLIC POLICY

10-13 APRIL 1996

Queen's University International Study Centre Herstmonceux Castle Hailsham, U.K.

LIST OF PARTICIPANTS

E.B. Andersen *Institute of Statistics, Denmark*
D.F. Andrews *University of Toronto*
B.A. Bailar *University of Chicago*
J.C. Bailar *University of Chicago*
G.A. Barnard *University of Essex*
D.J. Bartholomew *London School of Economics*
D.V. Bates *University of British Columbia*
T.A. Brzustowski *Natural Sciences and Engineering Research Council of Canada*
P. Calamai *The Otawa Citizen*
L.L. Conquest *University of Washington*
D. R. Cox *Nuffield College, Oxford*
S. R. Esterby *National Water Research Institute, Canada*
B. Farbey *University of Bath*
A.A. Farrant *The Acton Society Trust*
S. Fortier *Queen's University*
F.D.Gault *Statistics Canada*
E.A. Gehan *Georgetown University Medical Center*
J.F. Gentleman *Statistics Canada*
D.J. Hand *The Open University*
R.T. Haworth *Geological Survey of Canada*
J.R. Hemsley *Brameur/Vasari Enterprises Limited*
A.M. Herzberg *Queen's University*
C.C. Heyde *Australian National University*
G.N. Hood *Queen's University*
O.A.Y. Jackson *United Nations*
R.G. Jarrett *University of Adelaide*
J.F.C. Kingman *University of Bristol*
I. Krupka *Pest Management Regulatory Agency, Canada*
J.F. Lawless *University of Waterloo*
W. Leiss *Queen's University*
W.D. Macnamara *Queen's University*
D. McDonald *Monash University*
W.R. McGillivray *International Labour Organization*
J. Meisel *Queen's University*
P. Milliken *Member of Parliament, Canada*
G.H. Reynolds *Centers for Disease Control, Atlanta*
F.W. Scholz *Boeing Computer Services*
A.F.M. Smith *Imperial College of Science, Technology and Medicine*
T.M.F. Smith *University of Southampton*
R.G. Stanton *University of Manitoba*
A.T. Stewart *Queen's University*
R.E. Taylor *Stanford Linear Accelerator Center*
R.J. Tomkins *University of Regina*
D.H. Turpin *Queen's University*
D. Wroe *Central Statistical Office, U.K.*
H.P. Wynn *University of Warwick*
J.M. Ziman *Aylesbury, U.K.*